LIPIDS

Molecular Organization, Physical Functions and Technical Applications

Kåre Larsson

Chemical Centre, Box 124, Lund University, S-22100 Lund, Sweden

THE OILY PRESS

DUNDEE

ISBN 0 9514171 4 2

British Library Cataloguing-in-Publication Data. A catalogue record for this book is available from the British Library.

This is - Volume 5 in the Oily Press Lipid Library

(Volume 1 - "Gas Chromatography and Lipids" by William W. Christie; Volume 2 - Advances in Lipid Methodology - One" edited by William W. Christie; Volume 3 - "A Lipid Glossary" by Frank D. Gunstone and Bengt G. Herslöf; Advances in Lipid Methodology - Two" edited by William W. Christie)

Printed by Bell and Bain Ltd., Glasgow

To my family

PREFACE

The ambition behind this book is to describe the remarkable diversity of lipid structures as a basis for the understanding of lipid functionality in biological and technical systems. The common food processes involve the disruption of organized biological tissues, releasing the biomolecules to re-assemble themselves into arrangements which ultimately will become the final food product. Any scientific approach to improvement of food production must take the general structural principles and their underlying forces into account. As these changes involve the breaking and creation of interfaces, lipids play a crucial role, as they are the most surface-active of biomolecules. This knowledge is also required in the formulation of drug preparations to control targeting and release. Most industrial applications of lipids involve the unique and remarkable behaviour at a molecular level. Unless the fundamental properties of the molecules are understood, the only approach in lipid technology is trial-and-error. The disposition of the following text reflects this experience: the strong emphasis on molecular organization from which physical properties become obvious. Lipid molecular behaviour in all states of order in bulk phases and at interfaces is thus reviewed. The part of the book dealing with applications is focused on general phenomena, such as emulsification, and on food and pharmaceutical technology. During the writing of this book, several questions arose, which led to additional studies. New results are therefore presented, for example in the field of fat crystallization, transition mechanisms between cubic lipid-water phases, and phospholipids in cytoprotection.

Grants for my work in this field from Karlshamn AB is gratefully acknowledged. Dr W.W. Christie has given most valuable comments and improved my English.

Lund, November 1993
Kåre Larsson

CONTENTS

Part 2 Lipid Functionality

Part 1

Fundamental Aspects

CHAPTER 1

BASIC CONCEPTS

A. CLASSIFICATION OF LIPIDS

There is no strict definition of the term lipid that is generally accepted. A good definition was recently proposed (Christie, 1987): "Lipids are fatty acids and their derivatives, and substances related biosynthetically or functionally to these compounds". As physical behaviour is the main theme of this book it is natural from this perspective to divide lipids into two groups:

- *Polar lipids*, which interact with water giving aqueous phases, and
- *non-polar lipids*, which do not form lipid-water phases.

However, non-polar lipids can exhibit interaction with water at interfaces, and monolayers are thus formed on a water surface.

1. Non-polar lipids

General information on the chemistry of lipid molecules can be found in *The Lipid Handbook* (Gunstone *et al.*, 1986). Fatty acids can be saturated or unsaturated, and the predominant configuration of the double-bond in natural fatty acids is *cis*. The *trans*-isomer can be formed during hydrogenation; the most common modification of oils for commercial production of fats. The important group of non-polar lipids in foods is fats/oils (fats are partly solid). They are fatty acid esters of glycerol or *triacylglycerols*. We will use the older term, *triglycerides*, in this book. Because of variation in the fatty acid constituents, there is an enormous number of molecular combinations, and natural fats are therefore complex triglyceride mixtures. Butter fat, for example, can contain as many as a hundred thousand different triglyceride species.

Diglycerides behave like fats and oils from a physical point of view. This partial glyceride occurs in industrially processed lipids as an equilibrium mixture of similar amounts of the 1,2- and 1,3-diacylglycerols, as acyl migration takes place easily.

Sterols and their esters. Cholesterol is an important component of cell

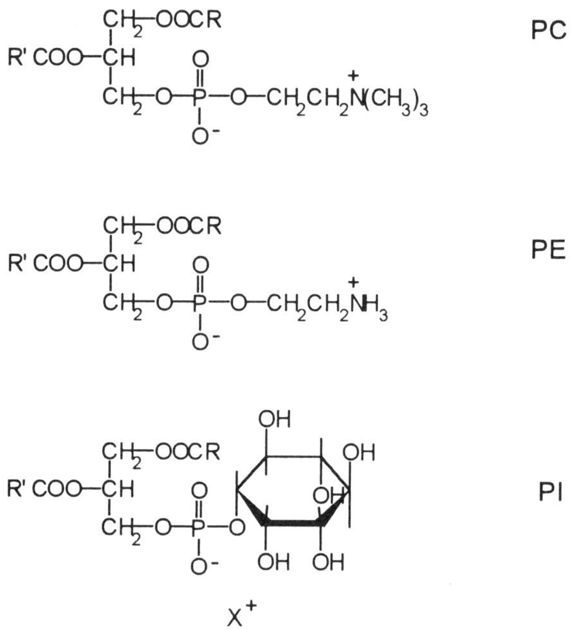

Fig. 1.1. Chemical formulae of phosphatidylcholine (PC), phosphatidylethanolamine (PE) and phosphatidylinositol (PI).

membranes in all tissues of animal origin, and it is also transported by lipoproteins in the blood as cholesterol esters. There are closely related sterols (such as stigmasterol) and esters in plants. Cholesterol is transformed into bile acids in the liver, and these are excreted into the intestinal system, where they act as detergents (and thus are classified as polar lipids).

2. *Polar lipids*

Monoglycerides are the simplest and probably also the most commonly used group of functional lipids. They are prepared industrially by molecular distillation, and an equilibrium mixture of about 90% of the 1-isomer and 10% of the 2-isomer is then obtained (*cf.* Krog and Lauridsen, 1976). This mixture behaves in a very similar way to the pure 1-isomer. The pure 2-isomer, which today can be prepared enzymatically to a reasonable purity, exhibits different physical properties as will be discussed in Chapter 3.

Phosphoglycerolipids are major cell membrane constituents, with acyl groups in the 1- and 2-positions and a polar head group involving phosphate in the 3-position of glycerol. The most important species, also from a technological point of view, is phosphatidylcholine (PC), shown in Figure 1.1. Other common phospholipids are phosphatidylethanolamine (PE) and phosphatidylinositol (PI).

Galactosylglycerolipids. The main origin of these lipids is the thylakoid membrane of chloroplasts, and they are therefore an important lipid class in all plants. In fact, they represent the most common polar lipid on the earth. The two major lipid classes in the thylakoid membrane are monogalactosyldiglycerides (MGDGs) and digalactosyldiglycerides (DGDGs).

Sphingolipids. These lipids contain a long-chain amine or sphingoid base, such as sphingosine, linked by an amide bond to a fatty acid. The sphingosine molecule in "free" form is charged at physiological pH, and it therefore behaves like a detergent. In addition, it contains two or three hydroxyl groups along the chain. The simple amide occurs, for example, in the skin (the lipid class is then termed a *ceramide*). Often, the terminal hydroxyl group on the sphingosine chain, next to the amine group, is linked to a polar head group. If this is phosphorylcholine (as in PC), the lipid is *sphingomyelin* (SM). The polar part can also consist of sugar groups (*cerebrosides* occurring in the brain, for example).

The first types of so-called *food emulsifier* were made industrially by adding glycerol to triglyceride oil and effecting inter-esterification. They were mixtures of mono-and diglycerides, and they have been used for more than half a century in the food industry.

Food emulsifiers are classified in the group of polar lipids. Mono- and diglycerides and phospholipids occur naturally in foods, whereas others are synthetic analogues of polar lipids.

The principle food emulsifiers used today are distilled monoglycerides and phospholipids. The common type of phospholipid is the crude mixture obtained on degumming of soybean or rapeseed oil. It is used under the name "lecithin" and contains many components beside phosphatidylcholine. Second in volume, among technical applications, is diacetyl-tartaric esters of monoglycerides (DATEMs), which are used in baking as "dough strengtheners". Next in importance are sodium or calcium salts of stearoyl lactylates (SSLs and CSLs respectively), which consist of a mixture of free fatty acids, and mono-, di- and even polylactoyl esters of stearic acid. The industrial production of these lipids has been described by Krog and Lauridsen (1989). In addition, food emulsifiers are used in pharmaceutical applications.

The monoglycerides which are used most often today are a relatively pure type prepared by molecular distillation and based on saturated acyl chains. Often, these are prepared from fully hardened lard. Because of acyl migration, the preparation process will yield the equilibrium mixture described above. The interaction with water and the physical properties of this lipid are very close to those of pure 1-monopalmitin (see Chapter 3).

There are other food emulsifiers used less frequently that are rather far from natural polar lipids in their chemical constitution. Examples are ethoxylated monoglycerides and polyoxyethylene sorbitan monostearate. They are very similar to synthetic surfactants from a chemical point of view, whereas those discussed earlier are much closer to natural lipids and their metabolites.

B. STRUCTURE, PHYSICAL PROPERTIES AND FUNCTIONAL PROPERTIES

In technical applications, lipids are frequently used to obtain specific functions. One example is emulsification of oil in water. In order to obtain an emulsion, we can add a polar lipid to an oil/water-mixture, and after a mechanical agitation process, an emulsion is obtained. The polar lipid used is a *functional* additive, and it is therefore necessary to discuss the *functionality* of the particular lipid used. Other functional properties involve flow behaviour or rheology. Thus, an oil becomes a plastic fat when we have added a critical amount of triglyceride crystals. Fat crystals as such are responsible for various functions in food emulsions, for example the possibility to whip an emulsion into cream and ultimately to form butter. Amylose complexing is an important method of modifying starch products, and this function is best achieved by polar lipids with one acyl chain only. Dispersion of sugar crystals in a fat continuum is another example where a polar lipid is needed for a specific function.

In Chapters 8 to 14, we will discuss such functions comprehensively and try to understand the underlying mechanisms. In order to do so we must understand how each functional property is related to the *physical* properties of our actual system. Emulsification, for example, is related to the viscosity of the oil and water phases, to the reduction of interfacial tension by the polar lipid used as emulsifier, to the mechanical properties of the interfacial film and finally to the kinetics involved in processing. The best approach to obtain a detailed knowledge of physical properties is to start from the molecular arrangements adopted by the lipid components, so that forces in different directions can be derived, for example. The combination of long-range order with short-range disorder is of utmost importance for most biological as well as technical functions of lipids, and these unique structural feature of lipids will therefore be treated in detail.

Lipid structure should be considered on two levels. Starting with the *molecular* level, lipid molecules have a remarkable ability to associate with other molecules of the same kind ("self-assembly") in different geometrical arrangements. In this way, *colloidal* aggregates are formed. A soap micelle is a typical example of a colloidal aggregate formed by polar lipid molecules. A particle in the size range 100 Å up to 10 000 Å is usually regarded as a colloid. It is obvious that a colloidal dispersion exhibits a large interfacial area towards the continuous phase. Dispersion of oil in water to form an emulsion, usually gives an oil-water (o/w) interfacial area of many tenths of m^2/g emulsion. The structure of the interface must play an important role for the external properties. This is the reason for the close link between surface chemistry and colloid chemistry. Surface and colloid science will be important tools when we try to link structure with functionality. The following chapters will focus on structure in all states of order. In connection with structure, we will also consider the effects on physical properties.

C. METHODS FOR CHARACTERIZATION OF STRUCTURE

In the solid state, X-ray diffraction is by far the most powerful method for structure determination. If single-crystals are available, the exact positions of each atom in the molecule can be determined, and such unambiguous data also provide a useful basis for the understanding of other states of order. The solid state behaviour of lipids discussed in Chapter 2 will involve description of the molecular packing in crystals.

The determination of structure from X-ray single-crystal data is today almost a routine task even for complex lipid molecules, and the bottleneck is the difficulty of growing perfect single-crystals. Earlier, computers set the limit for structure determinations; during the '50s it was only possible to obtain a projection picture of the crystal structure, and at the beginning of the '60s we needed a heavy atom in the molecule to derive the structure of large lipid molecules.

Even if single-crystals are not available, it is possible to derive a lot of information from X-ray powder diffraction data, which can always be recorded. By comparison with data from known structures, it is possible to determine hydrocarbon chain packing and the main orientational features of the molecules.

In the case of aqueous lipid-water phases, which are either liquids or liquid-crystalline phases, small-angle X-ray (SAX) methods are used to characterize the structure. These are related to powder diffraction, as no single-crystal domains are available. Small diffraction angles are necessary as the repetition distances are often very long. Neutron scattering methods are now applied to studies of lipid structures also. Replacement of hydrogen atoms by deuterium in lipids then provides a possibility of locating hydrogen atoms, a feature that is normally not possible by X-ray diffraction.

The frequency of X-ray (or neutron) radiation is so high in relation to the movement of the individual atoms that we can only "see" the time average of the atomic position. In order to obtain information on the dynamic behaviour of the molecules we must apply spectroscopy. The most powerful technique is NMR relaxation, a method that can provide further structural data. From infrared and Raman spectra we can obtain information on molecular conformation, for example whether the hydrocarbon chains of lipid crystals have their zigzag planes parallel or not.

The enormous increase in X-ray intensity achieved by synchrotron sources has recently made possible time-resolved X-ray diffraction analysis. This has been particularly important in order to follow the formation of lipid self-assembly in water. During the formation of liquid-crystalline phases, it is possible to record the diffusion coefficient of water also (Caffrey, 1991).

D. A MODEL LIPID

Monoglycerides of different chain lengths have been studied in the solid state, in aqueous systems and in monolayers. Therefore, they represent the polar lipid type which together with phosphatidylcholine is best known. Monoglycerides are furthermore the predominant functional additive in foods and they are frequently used in other applications, such as drug formulations. It has accordingly been natural to discuss monoglycerides in detail in this book, and to use them as a model representing polar lipids in general.

CHAPTER 2

LIPIDS IN THE SOLID STATE

A. CRYSTALLIZATION

In order to understand lipid crystallization, we will start with some general principles of formation of solids from liquids.

1. Crystal nucleation

The molecules in an oil are in chaotic movement, and this is determined by a compromise between entropy and adoption of effective packing arrangements. Well above the melting point, it is assumed that clusters of a few molecules are packed as in a crystal and that they are formed and dissociated dynamically. Such clusters will grow larger and larger with decreasing temperature, and ultimately a stable crystal nucleus is formed.

When a nucleus of radius r is formed, there is a free energy change ΔG, which is:

$$\Delta G = -4/3\ \pi r^3 \Delta G_v + 4\pi r^2 \sigma \qquad (1)$$

The first term is volume related and is due to the change in free energy at the transition from liquid to solid: ΔG_v. The second term gives the contribution due to the surface free energy, and σ is the specific interfacial free energy. The behaviour is illustrated in Figure 2.1. By derivation of this expression, we can get the maximum, which occurs at $r = 2\sigma/\Delta G_v$. This is the critical value of the radius for a nucleus to grow, or in other words the size above which a nucleus is stable.

If we consider triglyceride molecules, it is quite obvious that the radius is not an adequate measure of the size. It is more natural according to knowledge of crystal structure to imagine that "crystalline" dimers are aligned laterally to a bilayer unit of a certain size.

The nucleation rate is related to the free energy change at the critical nucleation dimension. Above a critical value of supercooling (or, for a solution, of supersaturation), the nucleation rate is so small that it can be neglected, and once this supercooling value has been reached the rate increases quickly (an exponential curve). The theory behind this is still not complete and is therefore not discussed further here. From van den

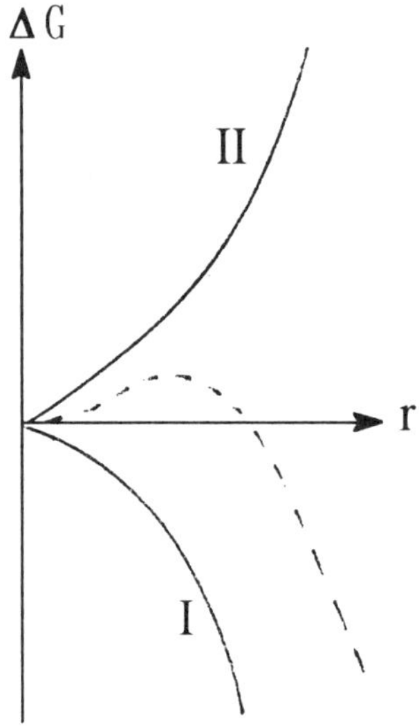

Fig. 2.1. The free energy change at nucleation is shown by the dotted curve, which is the sum of the volume term (curve I) and the interface energy term (curve II).

Tempel (1968), we obtain some relevant values for fats. Thus, a nucleus of main dimension r will be in equilibrium with a solution of concentration c_r at the same temperature at which a large crystal would be in equilibrium with a solution of lower concentration, if:

$$RT \ln c_r/c = 2 \sigma v / r \tag{2}$$

Or alternatively if we consider supercooling:

$$\Delta T = T_c - T = 2\sigma v T_c / r \Delta H \qquad 3)$$

The volume v of a triglyceride is about 900 cm^3/mole, the interfacial energy, σ, is about 10 ergs/cm^2, the heat of melting ΔH is about 1.4 x 10^{12} ergs/mole and RT is about 2.4 x 10^{10} ergs/mole. Using these values and a size of the nucleus (r) corresponding to ten molecules we obtain a supersaturation c_r/c of about 300 and a supercooling of 29°C.

There is an appreciable discrepancy between the supercooling value estimated above and the lower value normally observed. This shows that the nucleation process described here, that is *homogeneous nucleation* (or "spontaneous"), does not take place under the more usual conditions. Instead nucleation is induced by the presence of surfaces, which will induce orientation and order. This process is called *heterogeneous nucleation*, and the reduction in free energy in forming a nucleus of critical size (ΔG*) is then related to the contact angle between the liquid and the crystal nucleus on the actual solid surface, as shown in Figure 2.2.

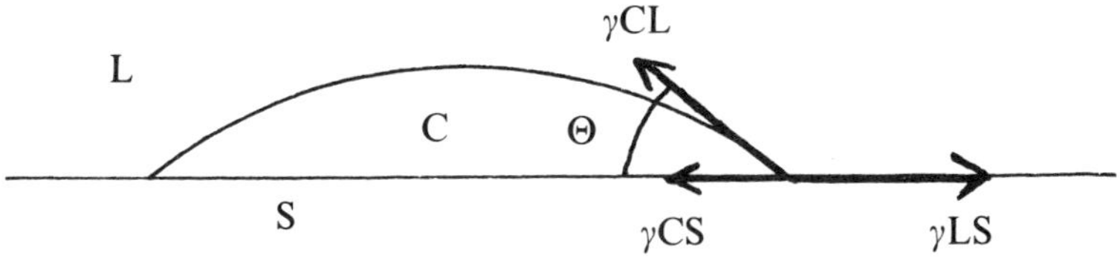

Fig. 2.2. Contact angle between crystal (C) and the liquid (L) on a solid substrate (S). Forces are balanced when $\gamma_{CS} + \gamma_{CL} \cos \theta = \gamma_{LS}$

The two nucleation processes described above are *primary* processes, and it is important to realise that *secondary nucleation* also takes place. Crystallization is usually performed in association with some agitation process, and it can then be seen how new crystal nuclei are formed on contact with existing crystals. The crystal surface will induce order in the contact layer of the liquid that results in new nuclei; this is secondary nucleation.

2. *Crystal growth*

When a crystal starts to grow, the difference in chemical potential of a molecule between being in the solution and in a site in the crystal will determine the growth rate. It is obvious that this difference is directly related to the undercooling of an oil or the supersaturation of a solution. If we consider the crystal structure, we can differentiate two growth situations. In the β-form of triglycerides, for example, we can imagine a very rough surface with many "boxes" of dimeric units missing from the outer surface (see Figure 2.21). The growth will be uniform or "continuous", when such a surface is filled up. The growth rate, v, for continuous growth is linearly related to the supersaturation (Garside 1987):

$$v = \text{constant} \times S \qquad (4)$$

where S is the supersaturation.

Once we have complete outer layers on such plate-like crystals, a two-dimensional nucleus must form on the surface, providing there are no defects. Depending on the detailed structure, there will be different contributions from the surface nucleation and the spreading of a layer, and the over-all rate has been proposed to follow the equation

$$v = A\, S^{P}\, e^{-B/S} \qquad (5)$$

where P is in the range -3/2 to 5/6 and A and B are constants (Garside, 1987).

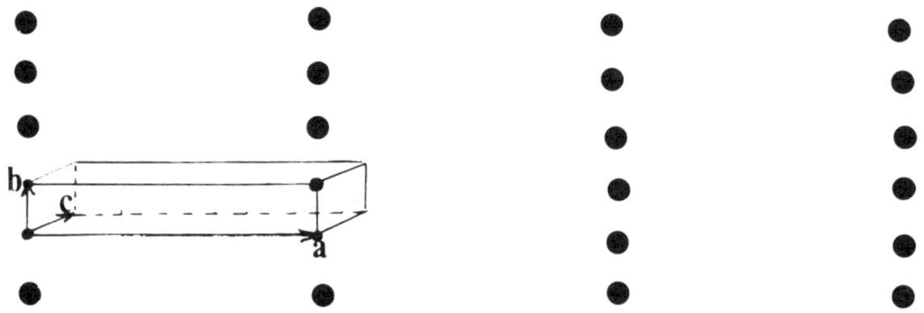

Fig. 2.3. Illustration of a unit cell of a crystal; the lattice in the a,b-plane is illustrated by points.

On the other hand, it has often been seen that crystal surfaces exhibit a spiral defect which can lead to continuous growth.

The growth unit must diffuse from the liquid to the crystal surface, and the concentration in a boundary layer will therefore be less than that in the bulk of the liquid. Under certain conditions the growth rate can therefore be diffusion controlled.

Additional features on crystal nucleation and growth, that are related to the particular structure of the triglyceride molecules, will be presented in the discussion of fat crystallization.

B GENERAL PACKING FEATURES OF LIPIDS IN CRYSTALS

1. Polymorphism

In the solid state, molecules adopt the ideal conformation and arrangement in relation to neighbours in order to optimise intra- as well as intermolecular interactions and achieve an efficient close-packing. The smallest building unit of a crystal, a box containing a few molecules, is called a *unit cell.* The crystal structure is obtained by repetition of this unit in its three axial directions: a, b and c as indicated in Figure 2.3. The repetition points of the unit cell constitute the *crystal lattice*, and crystallographic planes are various imagined planes through the lattice points. The lattice planes give diffraction by X-rays according to the well-known Bragg's law, and from the diffraction pattern the distances between the corresponding planes is obtained directly. This set of distances can be used to derive the unit cell, crystal class and symmetry (for example, whether two molecules in the unit cell are related by a mirror plane). The crystal classes and the corresponding dimensional conditions are:

- cubic: the three axes are equal and the angles between them, α, β and γ are all equal to 90°
- tetragonal: a = b and $\alpha = \beta = \gamma = 90°$
- hexagonal: a = b and $\gamma = 120$
- orthorhombic: $\alpha = \beta = \gamma = 90°$
- monoclinic: $\alpha = \gamma = 90°$

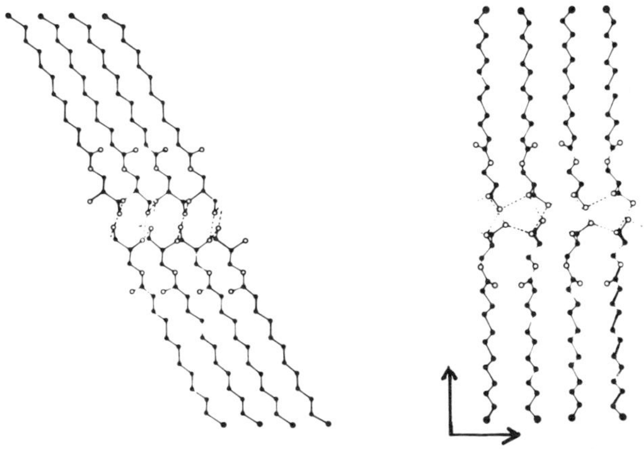

Fig. 2.4. Crystal structure of a 1-monoglyceride as seen projected along the shortest axis of the unit cell (Larsson, 1966a). Carbon atoms are indicated by filled circles and oxygen atoms by open circles.

- triclinic: no restrictions on axes ratios or angles.

If diffracted intensities from a single crystal are measured, it is possible to derive the atomic positions within the unit cell. These concepts are defined here because we will use them below in describing the structures.

The atomic positions in a crystal are all fixed, and there is only a certain degree of thermal vibration. A crystal structure characteristic for lipids is shown in Figure 2.4. The normal crystalline packing of lipid molecules is a stacking of lipid bilayers, that in fact reflects the amphiphilic nature of the molecule and a general organizational principle in other states of order also. By the parallel arrangement of the extended hydrocarbon chains in layers, the polar heads are exposed on surfaces, and this makes possible various kinds of polar forces between the atoms in the polar zone. As the detailed molecular organization down to the atomic level can only be determined in the solid state, it will be treated in considerable detail, as this knowledge is most useful in the understanding of structure and properties in other states of order.

Lipid molecules have the unique property of exhibiting different alternative packing in the solid state. This remarkable behaviour of simple lipids, *e.g.* in *n*-alcohols and fatty acids, as well as in complex lipids can be observed as phase transitions during storage. More spectacular, however, are multiple melting phenomena. During heating of the lipid, it first shows melting, then resolidification, before a new melting point at a higher temperature. In the case of complex triglycerides, such as the mixture of oleyl-distearins OSS:SOS in the ratio 1:1, slow heating after the first melting gives a liquid state lasting for hours before resolidification (Engström, 1992). Multiple melting in tristearin was first reported almost 150 years ago, but there was no explanation until 1934, when Clarkson and Malkin (1934) demonstrated by X-ray diffraction that different crystal forms existed.

n-Paraffins are sometimes regarded as the simplest type of lipid molecule. As long ago as 1930, Muller gave a thorough description of their structural changes in the solid state from very low temperatures to the melting point (Muller, 1930). He was thus able to describe their polymorphism and the molecular packing with the characteristic planar zigzag conformation of the hydrocarbon chain.

The occurrence of a periodicity along the chains equal to the zigzag period as well as the lateral close-packing geometry means that the chain packing is defined by a small cell within the true unit cell. We call this smaller unit a *subcell*.

The reason behind the polymorphic behaviour can be understood in the following way: There is only one best packing mode of molecules in the solid state. However, lipid molecules exhibit some order in the liquid state also, and during cooling of a melt, the molecule will first experience energy minima in nuclei with molecular conformations more related to those in the liquid. As such structures are metastable, further lowering of the temperature can give the activation energy for a phase transition into a more stable form. Certain triglycerides, like the most abundant species in cacao butter, show as many as six polymorphic crystal forms.

There are two main mechanisms behind polymorphism: The first is variations in the tilt of the molecules in the bilayers (see Figure 2.4). The second is the possibility of variations in hydrocarbon chain packing, which we will discuss specifically in the next section. In addition, there may be factors involved such as segregation of saturated chains in layers different from those of unsaturated chains. Such structures will be illustrated in the description of complex lipids at the end of this chapter.

The reason for the occurrence of different angles of tilt can be understood if we consider the molecular arrangement in Figure 2.4.We fix a molecule, for example one at the bottom to the left, and then we let the molecule located laterally on the right side slide one zigzag unit upwards, the next molecule in the right direction one zigzag unit upwards in relation to the new position of the molecule to the left, and so on. The molecules in the upper layer must of course be shifted in a similar way. The new structure obtained will look the same within the hydrocarbon chain region, but it will be different in the polar zone. When there are small polar head groups, like hydroxyl or carboxylic groups, it is possible to keep the same efficient hydrogen bonding systems along the polar sheets, and the energy difference between such differently tilted structures will be very small. Polymorphism in simple lipids can be explained in this way.

The physical properties of crystals of lipids with structure like that shown in Figure 2.4 reflect the layer arrangement. As there are very weak forces above the gap for the methyl end group, the structure can easily be opened up mechanically, for example towards an oil. The crystals work well as a solid lubricant. Furthermore the crystals can open up along the gap between the polar head groups towards polar media, and in Chapter 8 we will consider such surface activity of lipid crystals.

2. Hydrocarbon chain packing and nomenclature of lipid crystal forms

As the hydrocarbon chain packing arrangements are few and easy to identify by routine methods, it has been natural to base the nomenclature of the different polymorphic forms on chain packing geometry (Larsson, 1966c).

Simple lipids like alcohols, fatty esters and esters show three chain packing alternatives, which each can be described by the chain packing subcell (see Section B). When the melt of *n*-paraffins, alcohols or esters are cooled, a form with a hexagonal chain packing is first formed. The chains show rotational disorder along the hydrocarbon chain axes. Therefore, the packing is the same as that of cylinders. The rotation is reflected in the physical properties of the crystals; they are transparent and flexible. The term *wax* originally referred to these properties. By crystallization of *n*-nonadecane from chloroform, a film like ice on water is formed on the surface. This film made it possible to record X-ray single-crystal data, and the analysis confirmed free rotation of the hydrocarbon chains in the crystal lattice (Larsson, 1967a). The dimensions of the hexagonal unit cell are: a = 4.90 Å and c = 2.50 Å. All crystal forms with this chain packing are easily identified as they show one strong diffraction line at 4.15 Å in the short-spacing region. We call these crystal forms α-forms.

In complex lipids like phospholipids and triglycerides, there cannot be free rotation of the chains, as they are fixed at the polar head, but there is certainly a high degree of oscillatory mobility increasing towards the chain ends as indicated by spectroscopy data.

When an α-form is cooled we will ultimately reach a temperature when the rotational or oscillatory mobility stops, and there is a phase transition into a crystal form with fixed hydrocarbon chain planes. This transition is indicated in Figure 2.5, and the chain packing subcell obtained is orthorhombic. It is denoted $O\perp$ to indicate that every second zigzag plane is perpendicular to the rest. The dimensions of the subcell are a = 5.10, b = 7.40 and c = 2.55 Å. It gives two principle diffraction lines, at 4.2 Å and at 3.8 Å in the short-spacing region. We call these crystal forms β'-forms. Beside identification from the X-ray diffraction pattern, they can be identified by infrared spectroscopy, as they show a doublet at 720 cm^{-1} (the CH_2-rocking vibration region).

In the nomenclature conventions, we use the term β-form for all forms that are not α or β'-forms. There are many reasons for choosing this definition, and it agrees well with earlier terminology based on melting points. Furthermore, there are a few related chain packing subcells, which are hard to identify from their short-spacing diffraction, but in most cases, the most stable crystal form shows a triclinic subcell, T// (Figure 2.6), with

Fig. 2.5. Transition from the hexagonal chain packing cell (to the left) to the orthorhombic chain packing, O⊥. One zigzag unit of the hydrocarbon chain is shown projected along the chain direction (carbon atoms are shown by filled circles, hydrogens by open ones). The rings on the α-subcell indicate positions of carbon and hydrogen atoms in rotating chains.

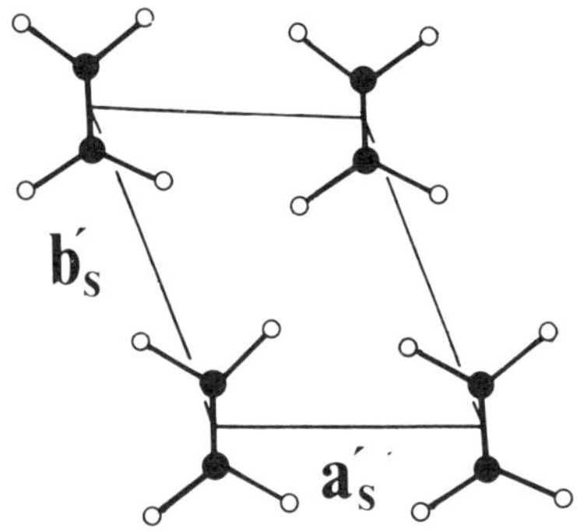

Fig. 2.6. The triclinic subcell T//, indicated as in Figure 2.5. The contents of the unit cell are projected on a plane perpendicular to the c-axis (therefore the use of a$'_s$ and b$'_s$).

dimensions a = 4.40 Å, b = 5.45 Å, c = 2.55 Å, α = 75, β = 10° and γ = 12° (adopted from trilaurin).

There are a few other chain packing alternatives observed in lipid crystal structures, although they are not as common as the three described above. An orthorhombic chain packing closely related to O⊥, called O′⊥, has been found, where all the chain planes are rotated 90°. Crystal forms showing this packing are evidently also β'-forms. A monoclinic subcell, M//,was observed in monoglycerides. All the chain planes are parallel, and it is as simple as the T//-subcell.

All chain packing arrangements represent the geometrical fitting of the hydrogen atoms, pointing outwards from one zigzag chain, into the grooves between hydrogen atoms in the neighbouring zigzag chain. This is easy to demonstrate by molecular models. There are a few more complex chain packing alternatives with more units in the subcell. The first (O//) was observed for a branched fatty acid, and the second (O′//) in crystals of oleic acid (Abrahamsson and Ryderstedt-Nahringbauer, 1962). Furthermore, complex hybrid cells, combining features from the simpler subcells, have been observed in crystal structures of membrane lipids (Pascher and Sundell, 1977).

3. Crystal Morphology and Equilibrium Shapes

When crystals grow very slowly, they will attain an equilibrium shape which minimises the interfacial energy between crystal and melt (or solution) of a particular crystal volume. As different crystal surfaces exhibit different interfacial energies, surfaces with low interfacial energy will grow at the expense of high energy surfaces. If we consider the methyl end group plane of a lipid crystal, it forms an interface towards methyl end groups in a disordered molecular layer of a melt. The interfacial energy is low, and this crystal surface will therefore always be a predominant one of lipid crystals.

Generally the equilibrium form of a crystal is the minimum of the sum over all crystal surfaces

$$\sum \gamma_i A_i$$

where γ_i is the surface free energy of surface i and A_i is its area. If we know γ_i and the crystal planes which form the crystal surfaces, this shape can be constructed by drawing normals to these planes from a centre point with a length proportional to γ_i. The planes are then drawn through these end points, giving the shape.

Usually, we do not know the surface free energy, although this knowledge is useful also in the reverse direction. From the crystal shape of a crystal grown very slowly, it is possible to estimate the relations between the surface free energy of different surfaces.

Because of differences in the surface free energy, large crystals will grow at the expense of small ones, and this is the mechanism behind the "ripening" process during storage of fat crystals in an oil continuum. Crystal surfaces are always crystallographic planes with high packing densities of atoms. An obvious restriction, however, is that the crystal surface forming a plane must be located between molecules. In a lipid crystal, the (001)-plane (parallel to the bilayer) will always form two dominant surfaces, as explained above. The other surfaces are usually parallel to the (001)-, (010)- or (110)-planes.

C. POLAR LIPIDS

1. Monoglycerides

In the sections above, we examined the crystal structure of a monoglyceride as an illustration of the principles of molecular organization of lipids. Racemic 1-monoglycerides of straight and saturated fatty acids exhibit four crystal forms. An α-form with hexagonal chain packing separates from the melt, and about 25°C below its crystallization temperature, a transition into the so-called sub-α form is obtained that

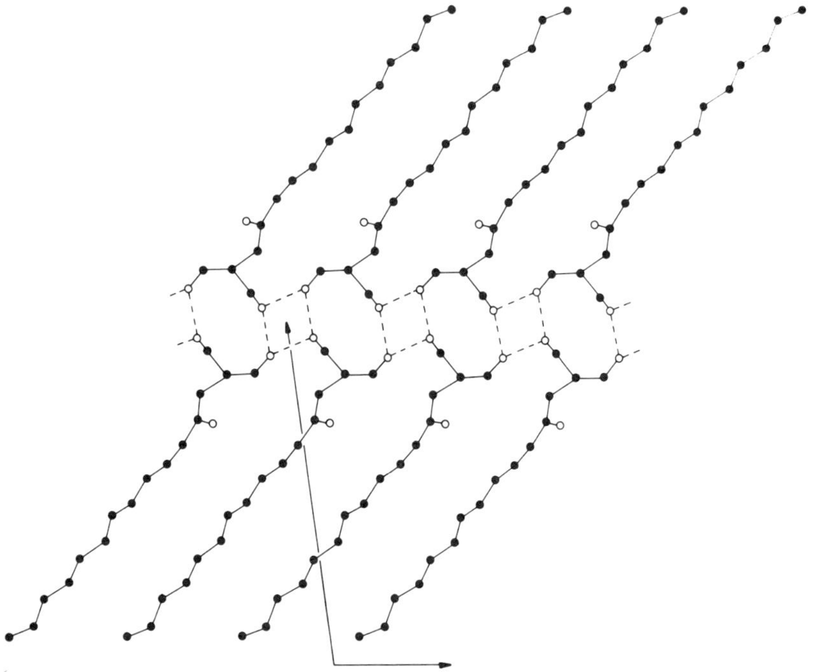

Fig. 2.7. Crystal structure of 2-monolaurin, after Larsson (1964a). The filled circles show positions of carbon atoms and open circles oxygen atoms.

involves a transition from the hexagonal to the orthorhombic chain packing mainly. From solution, the most stable form (β) is obtained, and it has a monoclinic chain packing subcell with all chain planes parallel. A metastable β'-form with orthorhombic chain packing can also be produced.

The most remarkable feature of these forms is that the bilayer thickness is the same in all (about 50 Å in the case of 1-monostearin), and this means that the chains have to adopt the same angle of tilt although the packing arrangement varies. This illustrates the dominating influence of the polar head group. We therefore have reasons to believe that the structure of the polar region is very similar in all forms (see Figure 2.4). Solvent crystallization can also give separation into crystals of each antipode. Polymorphism of 1-monoglycerides is therefore quite complex (Larsson, 1964b).

2-Monoglycerides of *n*-fatty acids show only one crystal form, which is shown in Figure 2.7. The angle of tilt is about 5° and higher than that of 1-monoglycerides. This is due to the configuration of the glycerol/acyl link. The glycerol group of the 1-isomer is just an elongation of the acyl chain, and this is not possible with the 2-isomer. This results in a larger cross-sectional area of the polar head groups at the bilayer surface, and a higher angle of chain tilt is required.

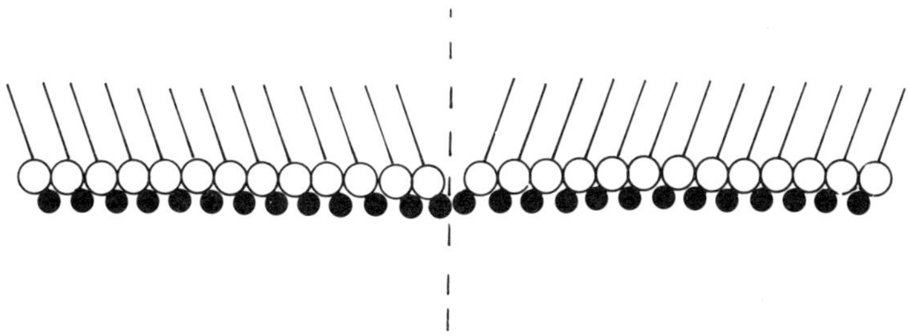

Fig. 2.8. Schematic illustration of the crystal structure in the A-form of potassium laurate, after Vand *et al.* (1949). Half of the bilayer unit is shown (filled circles denote potassium ions). On the left side, the tilt of a set of chains located in the same plane is shown, and on the right side of the dotted line, the chains in the adjacent plane are shown. The short axis is a and the long axis is c/2. The over-all bilayer structure is similar to that of other lipids, as illustrated in Figure 2.7, for example.

2. *Fatty Acid Salts*

Soaps are not easy to obtain as single crystals, and there are only a few crystal structures known. The first was the A-form of potassium caprate (nomenclature as for fatty acids). The structure is illustrated in Figure 2.8 (Vand *et al.* 1949). It is unusual in showing two directions of the hydrocarbon chains in the same monolayer.

This again illustrates the dominating influence of the polar head group on packing; the contributions from the hydrocarbon chains in the crystallization process play a minor role in relation to the ionic forces - they just have to adapt to the space available to them. By examination of molecular models, one can see directly that at a particular angle of tilt of one set of parallel chains, the hydrogen atoms of this chain layer fit well into the grooves of the oppositely tilted chains of an upper layer. We can describe the chain arrangement within each layer as crossed.

Only sodium and potassium soaps have been well studied, because of difficulties in growing single-crystals. The polymorphism in the solid state is complicated by incorporation of various amounts of water into the lattice, and successive changes towards liquid-crystalline phases take place. These structures will be discussed in the next chapter.

3. *Phospholipids*

Phospholipids are major components of cell membranes, and their structures are therefore of great biological importance. The first phospholipid structure reported was that of a PE, 1,2-dilauroyl-*rac*-glycerol-3-phosphorylethanolamine (Elder *et al.*, 1977). Soon afterwards the first PC-structure was reported, 1,2-dimyristoyl-L-glycerol-3-phosphorylcholine (Pearson and Pascher, 1979). They both form bilayer structures and the molecular conformation of each lipid is shown in Figure 2.9. The conformation of the glycerol backbone and the

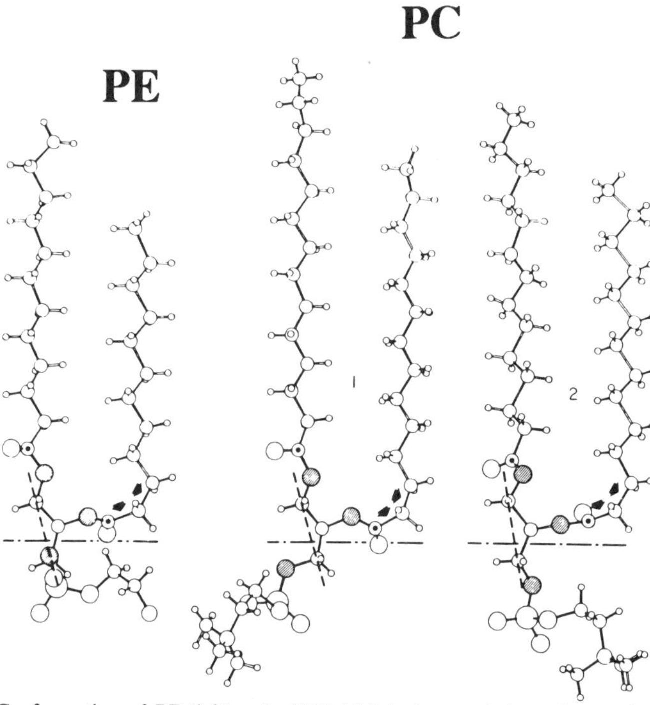

Fig. 2.9. Conformation of PE (left) and of PC (right), the two independent molecules in the asymmetric unit, after Pascher *et al.* (1981).

hydrocarbon chain attachment are very similar. More remarkable was the finding by Pascher and co-worker that the 1,2-dilauryl-L-glycerol showed the same conformation (Pascher *et al.*, 1981). This was taken as an indication of extensive conformational restrictions of the whole diglyceride region of the molecule, relevant also to the conformation in biomembranes.

Further studies showed that the glycerol group could have another conformation, however. The crystal structure of the sodium salt of dimyristoylphosphatidate showed a molecular conformation illustrated in Figure 2.10 (Harlos *et al.*, 1984). The glycerol group in PC and PE is an extension of the hydrocarbon chain direction (see Figure 2.9), whereas it is roughly parallel to the end-group plane in PA (Figure 2.10). It is most interesting to find that both of these alternatives also occur in triglycerides, as will be discussed in the section on fat crystallization (Section G). Similarly, the monoglyceride structure in Figure 2.4 shows the same glycerol group conformation as PE and PC.

Pascher and coworkers have determined the structure of other phospholipids. Dilauroyl-phosphatidyl-*N*,*N*-dimethylethanolamine, an intermediate compound in the stepwise methylation of PE to PC, shows a third conformation of the glycerol backbone (Pascher and Sundell, 1985).

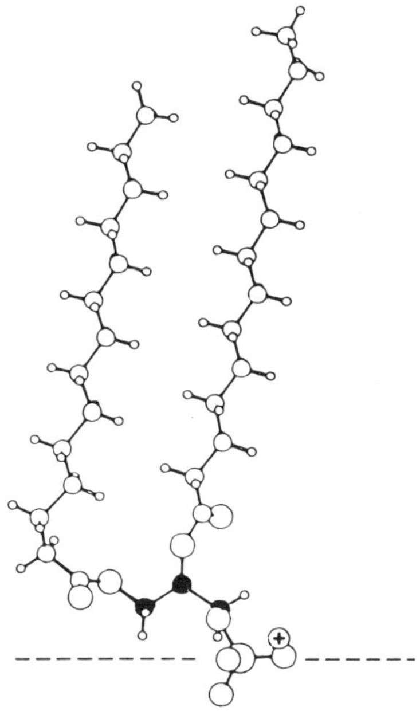

Fig. 2.10. Conformation of the phosphatidic acid molecule. The glycerol group is shown by filled circles and the end-group plane by a dotted line, after Harlos *et al.* (1984).

It represents an energy minimum when the chain-parallel conformation is rotated to form the layer-parallel one. Lysophospholipids were found to exhibit two types of structure. Lysophosphatidate (disodium salt) forms the usual bilayer type of structure, whereas a lyso-PC analogue, octadecyl-2-methyl-glycerophosphorylcholine shows a monolayer type of structure with interdigitating chains (Pascher and Sundell, 1985; Pascher *et al.*, 1986). The lyso-PC analogue structure is shown in Figure 2.11.

4. Glycosphingolipids

This group of lipids occurs in plasma membranes, and it is involved in intracellular recognition phenomena. A lyso-PC-analogue, glycosylphytosphingosine has a structure with interdigitating chains very similar to that shown in Figure 2.11 (Abrahamsson *et al.*, 1977). The structure of a cerebroside, with two chains as in phospholipids, has been determined (Pascher and Sundell, 1977). It forms a bilayer structure with the galactosyl group bent in relation to the chains, so that the molecules have the shape of shovels.

D. NON-POLAR LIPIDS

1. Alcohols, Fatty Acids and Simple Esters

These lipids are classified as non-polar lipids, as they do not form aqueous phases, in accordance with the definition given in Chapter 1. On

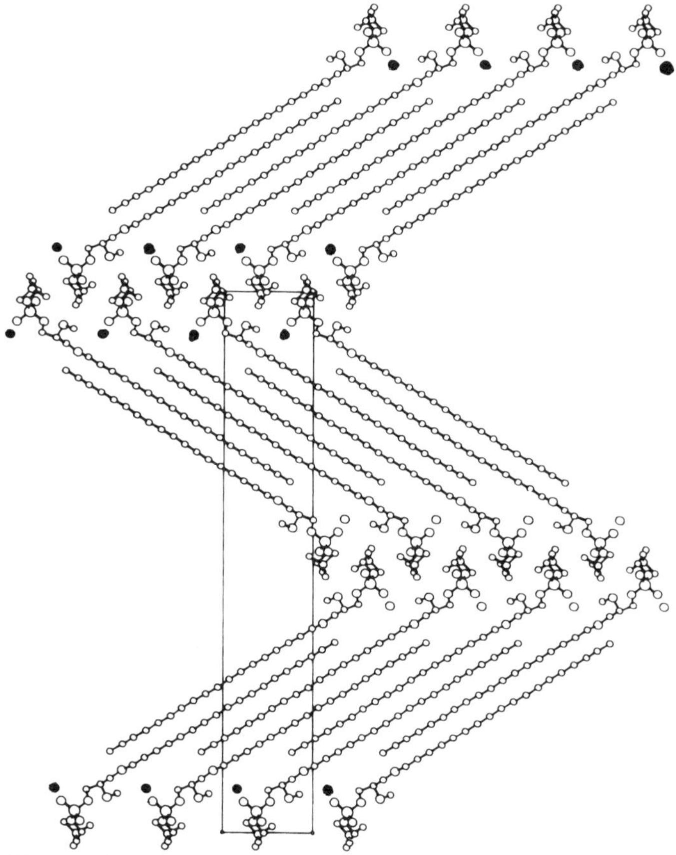

Fig. 2.11. The crystal structure of octadecyl-2-methyl-glycerophosphorycholine, after Pascher *et al.* (1986).

the other hand, it should be realised that they still have a polar group, which will determine the solid state packing in much the same way as we have seen for polar lipids.

Long chain *alcohols* crystallize in a vertical α-form, and this means that the molecules of the bilayer are disordered in a manner corresponding to rotation along the chain axes. Members with an even number of carbon atoms have three forms. One crystal form has a structure similar to that of the α-form, with the chains frozen into the orthorhombic chain packing. The third form has tilted chains in the orthorhombic chain packing. This structure is shown in Figure 2.12 (Abrahamsson *et al.*, 1960). Odd members also show three crystal forms, all vertical.

Fatty acids have been studied extensively since the 1920s. The nomenclature of fatty acid structures has been based on the X-ray long-spacing, *i.e.* the angle of tilt of the molecules towards the bilayer plane. In the order of increasing tilt, they are termed A, B and C in the case of fatty

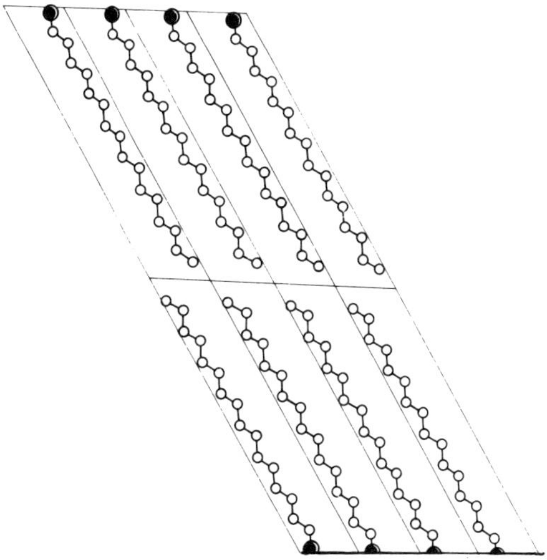

Fig. 2.12. Crystal structure of n-hexadecanol, after Abrahamsson *et al.* (1960).

acids with an even number of carbon atoms, and A′, B′ and C′ in the case of odd-numbered acids. The even and odd series are similar in chain packing; A and A′ have the triclinic packing T//, whereas all the others have the orthorhombic packing O⊥. The general features as described by von Sydow (1955) are as follows. The A-form has the molecules arranged laterally head-to-tail, so that the structural unit is a monolayer. There is a superstructure alternative where groups of three molecules are packed head-to-tail in the monolayer. The two structures are shown in Figure 2.13. The A′-form exhibits the usual bilayer structure with a tilt of about 7°. In contrast, the B-form is a bilayer structure with a most remarkable bend of the hydrocarbon chain at carbon atom 2 (gauche-conformation between atoms 2 and 3 from the carboxyl group carbon), as determined by Goto and Asada (1978). The B′-form is a bilayer structure. Similarly, the C and C′-forms are bilayers, as illustrated in Figure 2.14. The tilt is about 8°.

The polymorphic changes of even-membered homologues of fatty acids can be described by the sequence observed at heating : A → B → C → liquid (von Sydow, 1955). All three forms can be crystallized from solvents under varying conditions, whereas only the C-form can be obtained from the melt. The irreversible transitions to the C-form show that it is the stable form. Odd members exhibit similar transition relationships, with the difference that B′ can also be obtained from the melt.

Among unsaturated fatty acids, the structure of oleic acid was determined first (Abrahamsson and Ryderstedt-Nahringbauer, 1962). The molecules form bilayers with tilted hydrocarbon chains, and at the *cis*-double bond there is a change in the direction of tilt. The crystal structure of linoleic acid has also been determined, and it is quite analogous, with

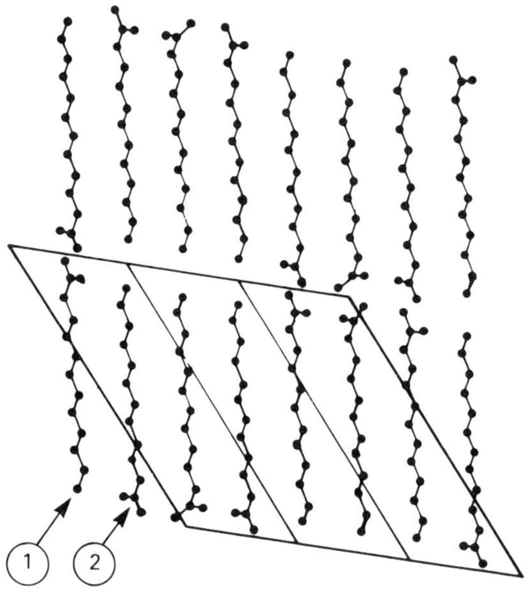

Fig. 2.13. Crystal structures of the A- and A-super forms. That of the A-form (molecules 1 and 2 in the unit cell) illustrated *within* the superstructure of dodecanoic acid, after Larsson (1986b).

two changes in tilt direction (towards the opposite tilt direction at each *cis*-double bond) (Ernst *et al.*, 1979). The structure is shown in Figure 2.15.

The polymorphic behaviour of oleic acid has been analysed in detail by Suzuki *et al.* (1985). They identified three forms, α, β and γ (using the nomenclature described in Section B).

Various binary systems of fatty acids have been studied, and phase properties of the unsaturated ones are most complex (*cf.* Yoshimoto *et al.*, 1991).

Simple esters of fatty acids with a short-chain alcohol have been well characterised, but it has not yet been possible to grow single-crystals from fatty acid esters of long-chain alcohols. Methyl stearate gives two β-forms from the melt as well as from solvents (Aleby, 1969). The most stable form, β', is obtained on heating of the β'_2-form at 38°C. The crystal structure of methyl stearate in the β'-form looks just like those of other lipids discussed above, with a lipid bilayer where the polar ester group region is arranged head-to-head.

In contrast, ethyl stearate, like higher esters, is arranged as shown in Figure 2.16. This illustrates that the polar group is too far from the end of the molecule to influence the orientation of polar groups in neighbouring layers. The molecules of adjacent layers are thus arranged head-to-tail. A remarkable feature is that the polymorphic behaviour is similar to that of

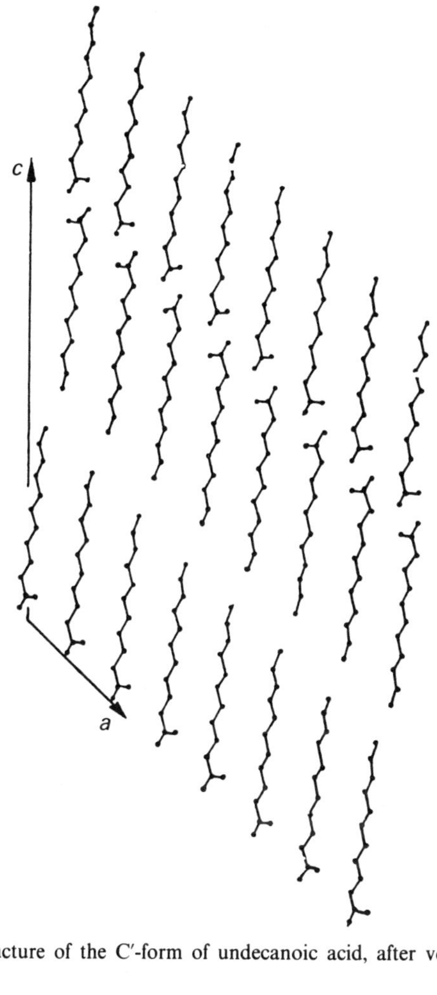

Fig. 2.14. Crystal structure of the C′-form of undecanoic acid, after von Sydow (1955).

methyl stearate, although the molecular orientations in the crystals are different (Aleby, 1968).

Fatty acid esters of long-chain alcohols occur frequently in plants, for example as a protective coating (cuticle) on leaves and fruits. As sperm whale oil contains about 76% of wax esters, it has unique properties, and this was one reason for extensive hunting of whales at the beginning of this century. A natural source of similar esters used today is the *Jojoba* plant. Long-chain alcohol esters have a tendency to give quite stable α-forms on crystallization, and the term wax is related to the plastic properties of this crystal form.

In mammals, cholesterol is the most common natural alcohol component of simple esters. Cholesterol esters are major lipid components in blood lipoproteins. Their packing arrangement is related to the acyl

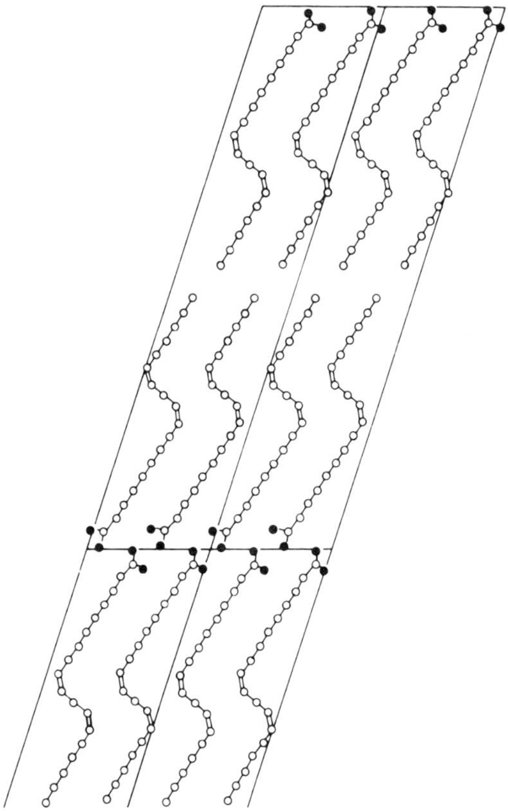

Fig. 2.15. Crystal structure of linoleic acid, after Ernst *et al.*(1979).

chain length. With longer fatty acids, such as in cholesteryl myristate, the cholesterol ring systems form a centre bilayer with the hydrocarbon chains pointing outwards (Craven and De Tita, 1976). By interpenetration of hydrocarbon chains, one cholesterol skeleton gets the same cross-sectional area as two hydrocarbon chains, as shown in Figure 2.17.

The structure of cholesterol oleate contains a monolayer of the cholesterol ring systems in the centre of the unit layer (Craven and Guerina, 1979). By a high angle of tilt of the interpenetrating chains, the cross-sectional area of one cholesterol skeleton corresponds to that of one chain, as illustrated in Figure 2.17. A further remarkable feature is that the *cis*-double bond only gives a minor kink in the direction of the chain.

There is a third arrangement in esters of C_9, C_{10} and C_{11} chains, where cholesterol ring systems and hydrocarbon chains are arranged in the same layers. These medium-chain cholesterol esters form remarkable liquid-crystalline phases, which will be discussed in the next chapter.

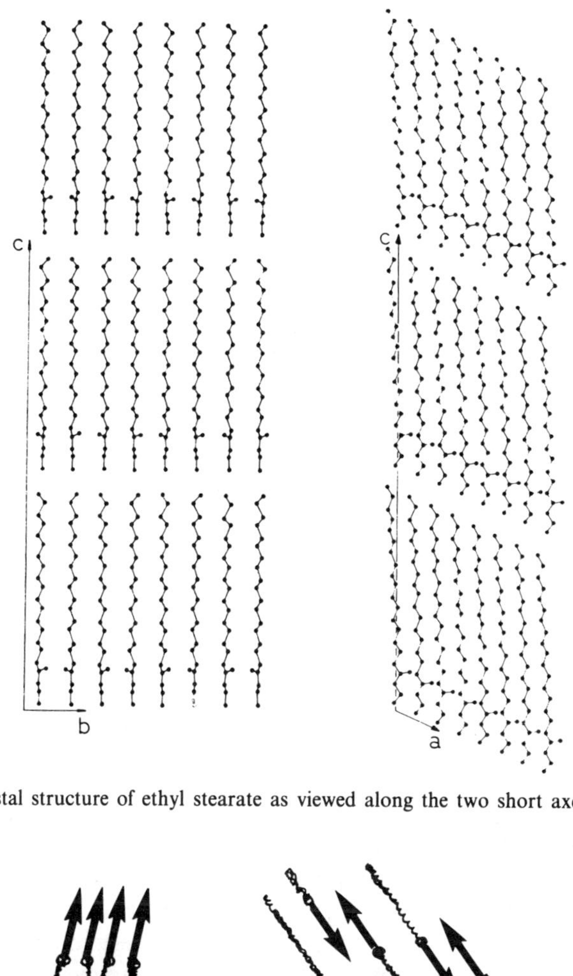

Fig. 2.16. Crystal structure of ethyl stearate as viewed along the two short axes, after Aleby (1969).

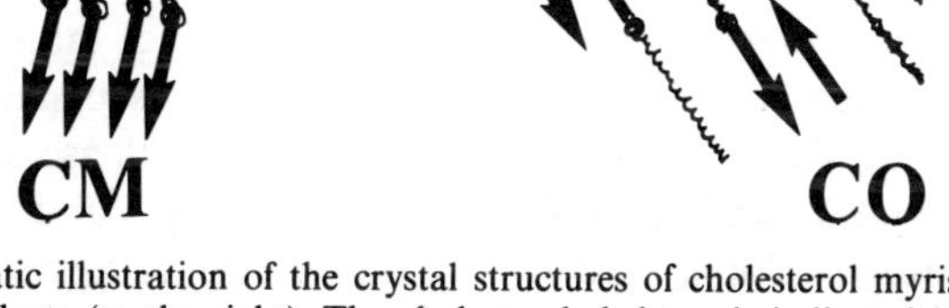

Fig. 2.17. Schematic illustration of the crystal structures of cholesterol myristate (to the left) and cholesterol oleate (to the right). The cholesterol skeleton is indicated by an arrow, and the chain axis is also shown.

2. *Non-Polar Membrane Lipids: Cholesterol and Diglycerides*

Cholesterol can crystallize in human tissues under pathological conditions in the form of a monohydrate, which is easily transformed into anhydrous crystals (Bogren and Larsson, 1963). The anhydrous form has

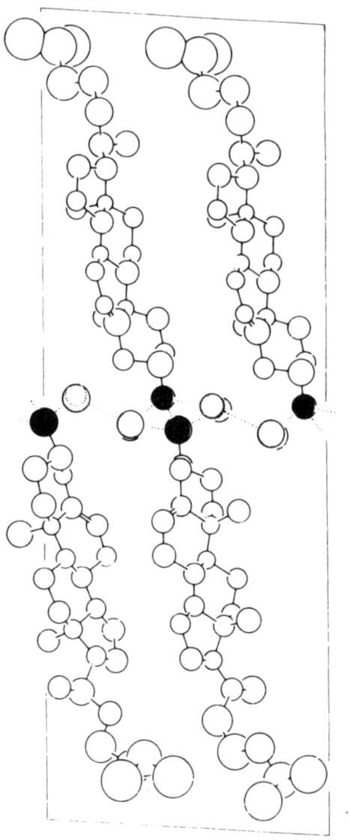

Fig. 2.18. Crystal structure of cholesterol monohydrate, after Craven (1976).

two polymorphic crystal structures, and both have been determined (Hsu and Nordman, 1983). They differ from the usual lipid bilayer structure as there is not a planar polar sheet. Hydrogen bond-associated dimers are arranged in pairs, and adjacent groups are displaced, so that the bilayer has a rippled shape. The crystal structure of cholesterol monohydrate is shown in Figure 2.18 (Craven, 1976). It is of the characteristic lipid bilayer type with hydroxyl groups and water molecules forming a hydrogen bonded sheet in the centre of the bilayer.

Cholesterol monohydrate exhibits an interesting transition into a smectic state (a lamellar type of liquid crystal discussed in the next chapter) at 123°C. The X-ray diffraction data of this phase show a bilayer thickness of 37.4 Å and a diffuse short-spacing band, indicating a disordered lateral packing. It can contain one to four water molecules. This smectic state was found and characterized by Loomis and coworkers (1979).

Diglycerides occur as components of transitory existence in cell membranes, where they function as second messengers and are formed by enzymatic degradation of phosphatidylinositol phosphates. Diglycerides can occur in two types of crystal form. Structures of two 1,3-diglycerides have been determined, and that of 3-thiododecanoic acid is shown in

Figure 2.19. The structure of the 1,3-diglyceride of 11-bromoundecanoic acid has the same V-shaped molecules (Hybl and Dorset, 1971). Thus, the chains are extended with the polar region forming the centre of the unit layer, which is a monolayer. Subsequently, the structure of a 1,2-diglyceride, 1,2-dilauryl-L-glycerol, was determined (Pascher *et al.*, 1981). In this instance, the molecules have both chains pointing in the same direction, so that the structure is of the usual bilayer type. This diglyceride type can occur in cell membranes, and the crystal structure shows the probable conformation in the bilayer.

E. FAT CRYSTALLIZATION

The properties of the group of non-polar lipids known as fats are very important in food technology, and their solid state behaviour, which is treated below, is especially relevant.

1. Simple Triglycerides

In using the term "simple", we mean that they contain only one type of fatty acid, and we will first consider *n*-saturated ones.

During many decades, fat polymorphism, particularly the number of crystal forms of simple triglycerides, was a controversial field. One probable explanation is the poor quality of triglyceride samples available at that time. Much of the confusion was due to the description of a vitreous form, which clearly does not exist. Detailed studies based on infrared spectroscopy (Chapman, 1962) and X-ray diffraction (Larsson, 1964b) resolved this controversy and gave consistent results on the polymorphic behaviour. The transitions are indicated in Figure 2.20.

The melt crystallizes into an α-form on cooling, and by rapid cooling to very low temperatures (about -20°C at chain lengths of C_{18}), there is a transition to a sub-α form (which is a β'-form according to the nomenclature defined in section B, but we prefer to call it sub-α because of its intimate relation with the α-form).

When the α-form is stored just below its melting point, it is transformed first to the metastable β'-form, and finally into the stable β-form. The melting point of the β-form is about 10°C higher than that of the α-form, and the β' exhibits an intermediate value. The stability of the α-form increases with chain length; tristearin, for example, is quite stable at room temperature. The stability of this form also increases with the presence of impurities, since molecular species of other chain lengths can co-crystallize into the α-lattice.

2. Crystal Structures

In 1963, the crystal structure of the β-form was determined independently by Jensen & Mabis and by Larsson. The triglycerides used

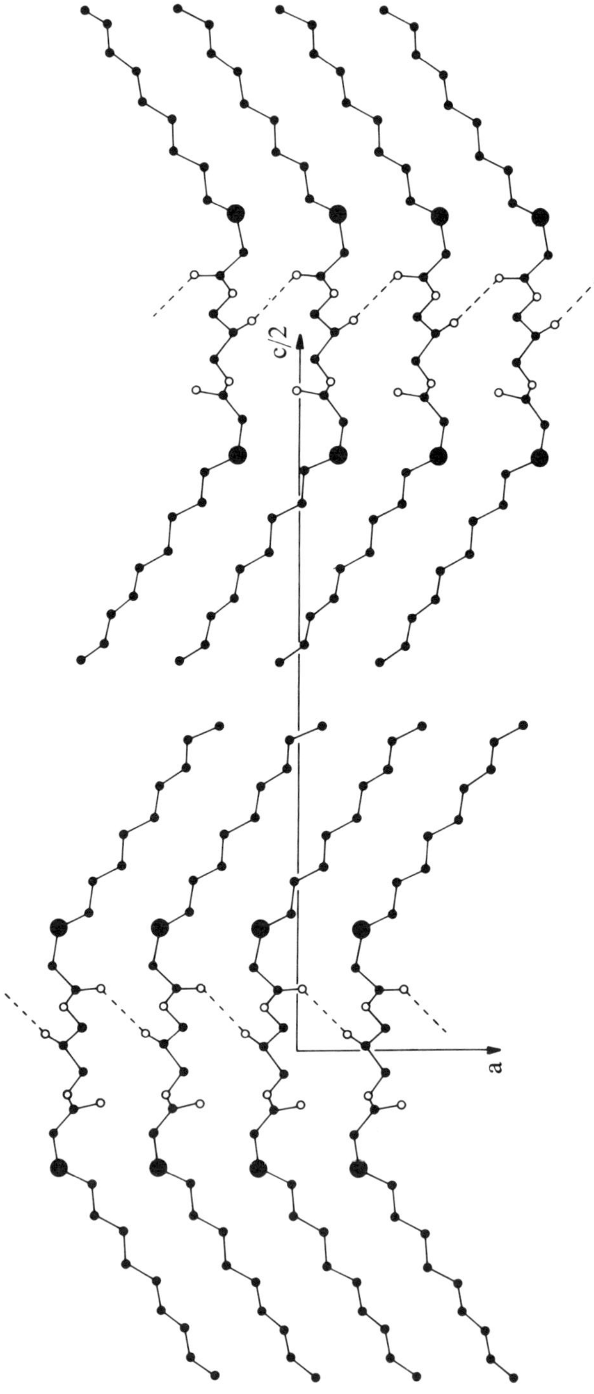

Fig. 2.19. Crystal structure of the 1,3-diglyceride of 3-thiododecanoic acid (Larsson, 1963a). The dotted lines show the hydrogen bond system.

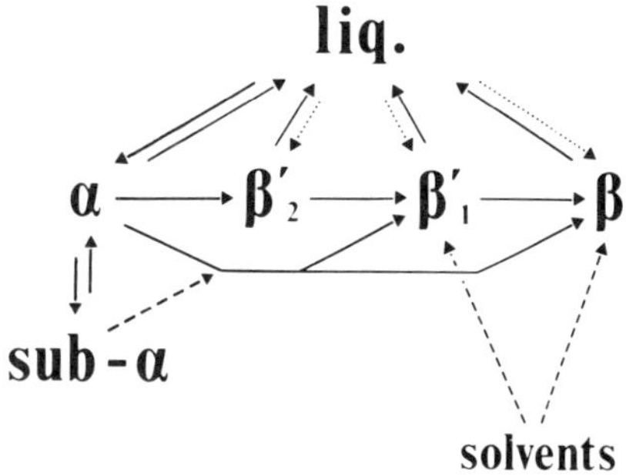

Fig. 2.20. Polymorphic transition in saturated monoacid triglycerides. Dotted lines indicate that the transition only occurs under special conditions of thermal treatment.

were tricaprin (Jensen and Mabis, 1963) and tri-11-bromoundecanoin (Larsson, 1963b) with its isomorphous form trilaurin (Larsson, 1964a). The crystal structure of trilaurin is shown in Figure 2.21. The glycerol group allows the chains in the 1- and 2-positions to form an extended chain of which the two glycerol carbons are a part. The chain in the 3-position forms a branch aligned in parallel with this long chain. By arranging two such molecules in pairs in opposite directions, a very effective space-filling arrangement is obtained. The hydrocarbon chains are close-packed according to the triclinic subcell. The methyl end groups cannot be located in one plane, as in lipids with one chain only (*cf.* Figure 2.4). A rippled end-group structure is therefore obtained. The geometry of these ripples gives restrictions on possible variations of chains within the molecules, as will be discussed later.

The structure of the β'-form has not been determined completely. It forms thin needles with a remarkable tendency for twin growth, making the determination of single-crystal structure complicated. The main features of the structures have been derived in the case of triundecanoin, however (Hernqvist and Larsson, 1982). The structure is shown in Figure 2.22. The conformation of the polar head group could not be derived. However, it must be closely similar to that of the β-form, shown in Figure 2.21, as the geometry in the glycerol group region is almost identical. The chains are packed according to the orthorhombic chain packing subcell.

It should be mentioned that tilted molecules normally show an alternation of melting points between even and odd members of a homologous series. This is rather natural if we consider a methyl end group plane like that in the β-form, for example (see Figure 2.21). Elongation with one CH_2-group of the chains will give a different appearance of the methyl end group planes. This will result in a different lattice energy and therefore different melting curves of homologues of molecules varying in chain length. Part of the confusion about triglyceride polymorphism mentioned earlier was related to the observed non-alternating melting point of the β'-form, and it was therefore assumed to

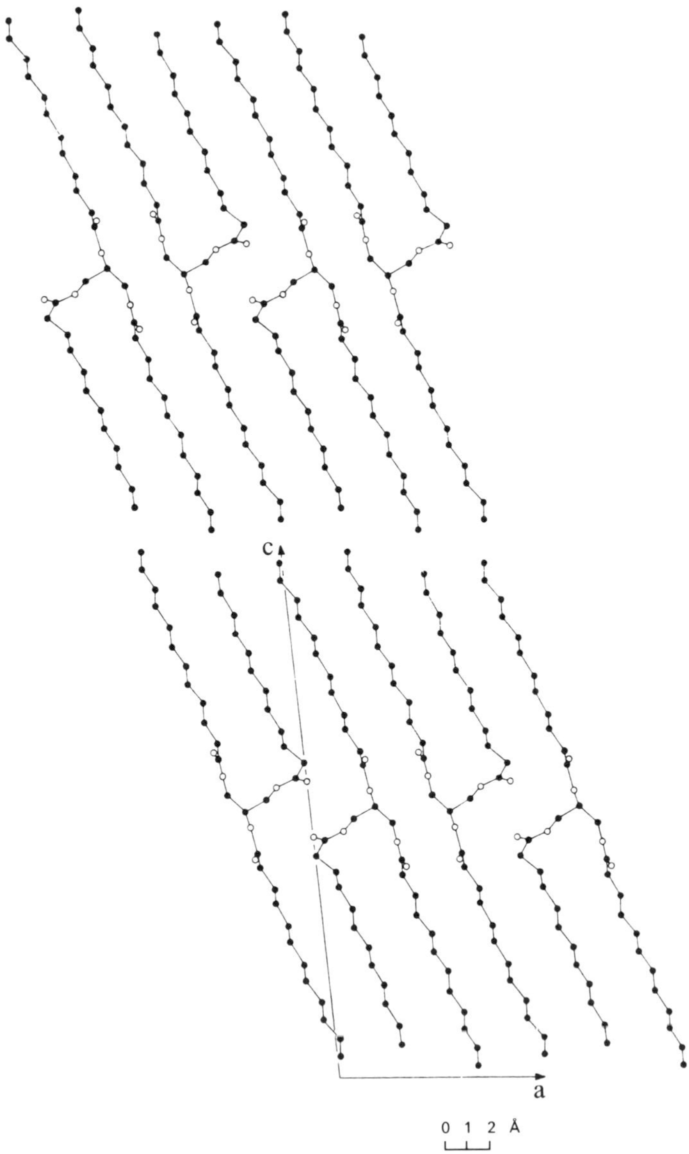

Fig. 2.21. Crystal structure of the β-form of trilaurin as viewed along the shortest axis of the unit cell (atoms are denoted as in Figure 2.4)

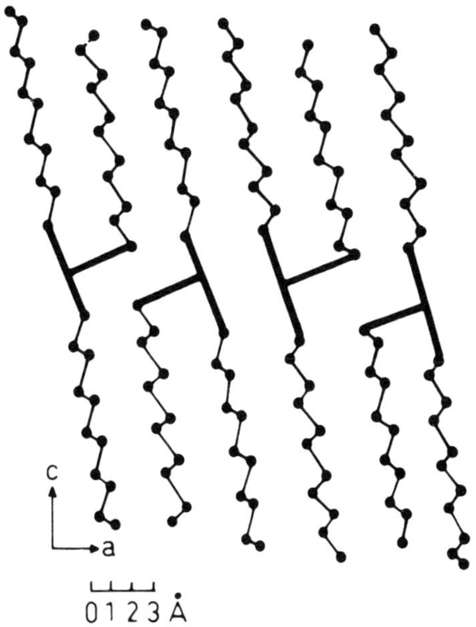

Fig. 2.22. Crystal structure of the β'-form of triundecanoin (atoms are denoted as in Figure 2.4) (Hernqvist and Larsson, 1982). The unknown glycerol group region is indicated by the molecular backbone.

have vertical chains in relation to the end-group planes. On the other hand, an analysis of the end-group structure showed that with the specific orientation of the end group plane in β' in relation to the chain packing subcell, the even and odd members give similar end group structures. No alternation should therefore be expected (Larsson, 1966b).

There is still an unresolved question concerning the existence of a second β'-form. With odd members from C_{11} to C_{17}, we found two different β'-forms (Hernqvist and Larsson, 1982). At the same time, Simpson and Hagemann (1982) described two β'-forms of tristearin, and it therefore seemed natural to assume that two β'-forms always exist. A recent study of tripalmitin shows that there is no strict crystallographic difference between the two forms, however (Kellens *et al.*, 1990). They seem to differ in degree of crystal perfection only. From this study, it seems likely that in even triglycerides there is only one β'-form. In odd numbers, there are clearly two forms, and X-ray data indicate that they differ in the orientation of the orthorhombic chain packing subcell in relation to the end-group plane (Hernqvist and Larsson, 1982).

In mixed-chain triglycerides, there appears to be a β'-form with a somewhat different crystal structure (Birker *et al.*, 1991) (see also next section). It has roughly the same unit cell dimensions as β'-triundecanoin and also alternation of the tilt angle between adjacent layers; the remarkable difference is that the chains in the 1- and 2-positions point in the same direction.

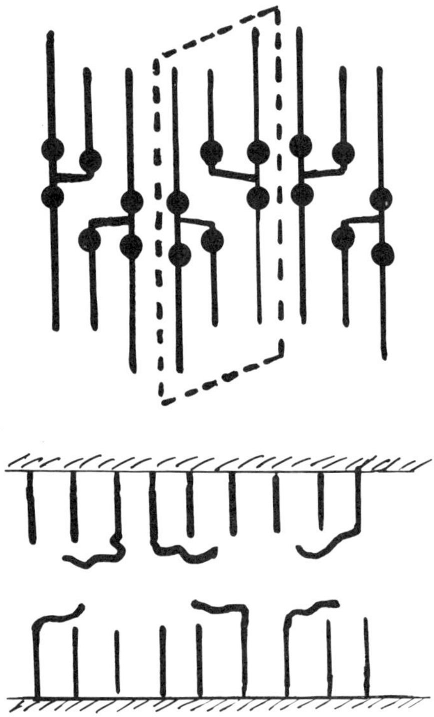

Fig. 2.23. Proposed α-structure in simple triglycerides (Larsson, 1992b). The lateral arrangement in bilayers is shown above, with the dimeric unit (dotted) related to adjacent units by mirror-symmetry. The proposed disorder of the chains at the methyl end group plane is shown below.

We have now the structure of the α-form left to discuss. The only features we can derive from the X-ray powder data is the chain packing, according to the hexagonal subcell, and that the molecules are organized in a bilayer, *i.e.* in a double-chain-length structure. The α-form has vertical chains in relation to the end group plane, and the fact that the dimensions of the glycerol group region of the α-form compared to the β′- and β-forms of simple triglycerides are quite similar according to X-ray data indicates that the glycerol residues of the α-form have the same main structure. Based on this assumption, the proposed structure of the α-form is indicated in Figure 2.23. Thus, the vertical hydrocarbon-chain arrangement should be expected to require a mirror-plane symmetry of the dimeric unit (indicated by broken lines in Figure 2.23). Any stacking of such bilayers must contain disordered end-group regions at the methyl end group gap, and one such alternative is indicated in Figure 2.23. This type of disordered structure can accommodate various triglyceride molecules with different chains in the same crystal; co-crystallization and solid solution formation is in agreement with this, and is much more common in the α-form in comparison to β′- and β-forms.

3. Mixed-Chain Triglycerides

Crystal structures of *mixed triglycerides* have been determined in two cases. The symmetrical triglyceride with capric acid in the 1- and 3-positions and 11-bromoundecanoic acid in the 2-position (the latter should be expected to crystallize like lauric acid) was found to crystallize like simple monoacid triglycerides (Doyne and Gordon, 1968). Furthermore, the crystal structure of 1,2-dipalmitoyl-3-acetyl-*sn*-glycerol has recently been reported (Goto *et al.*, 1992). The molecules form a trilayer structure with triclinic chain packing in the palmitic acid bilayers; thus it is a β-form. The short acetyl chains interdigitate in one layer, and the structure is discussed below.

In addition, the main features of the structure of the β'-form of 1,3-dilauroyl-2-myristoylglycerol have been determined by Birker *et al.* (1991). They recorded single-crystal data, and by comparison with powder X-ray data from molecular species of related mixed triglycerides of myristic acid (M), palmitic acid (P) and stearic acis (St) (like PStM/MPP, PStL/LMP and PStC/CLP) they were able to prove that the β'-form of triglycerides of the $C_nC_{n+2}C_n$-series adopt a conformation where chains in the 1- and 2-positions are directed in the same way in relation to the glycerol group.

It may be possible to generalise mixed triglyceride structures on the basis of the two different glycerol backbone conformations in monoacid triglycerides and in dipalmitoylacetylglycerol. The four different structural alternatives that seem possible in triglycerides are shown in Figure 2.24. Either the chain in the 2-position is directed opposite to that in the 1- and 3-positions (as in simple triglycerides), or the chain in the 2-position is located in the same layer as the 1- or 3-chain (as in dipalmitoylacetylglycerol). These are the two basic conformations defining the directions of the chains. The second level of structure is determined by the similarity of the hydrocarbon chains. If the three chains are rather similar in length or unsaturation, the bilayer conformation (or double chain-length structure) is obtained, where all three chains of the molecule are mixed in the same chain monolayer. The alternative shown in Figure 2.24 is formed when one chain of the molecule segregates into one chain layer and the remaining two chains are located in another layer. This alternative gives rise to a triple-chain length structure.

On a higher level of structural detail, the chain direction may alternate between hydrocarbon chain layers, and even the chain packing can vary.

When a melt is cooled, there is a tendency to keep the bilayer structure of the liquid state (see below) and to form the hexagonal chain packing first. Once this α-form has crystallized, the structures of the more stable forms obtained later will be determined mainly by the tendency for equal hydrocarbon chains to be in the same chain layer. If one chain differs much from the other two in length or in unsaturation, segregation will occur, so that a triple-chain-length structure will be formed. There is an

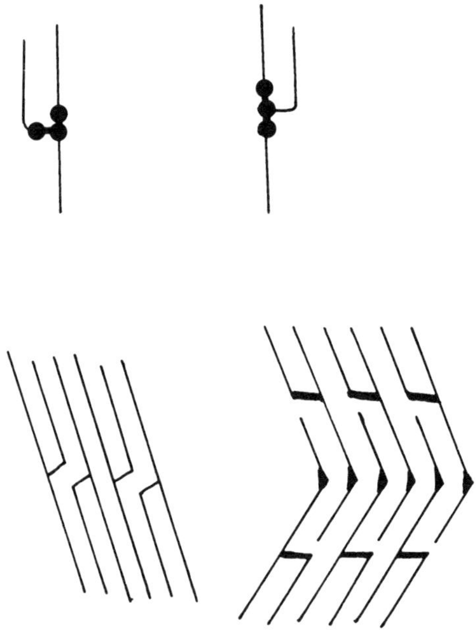

Fig. 2.24. Triglyceride conformation types in crystals. The two chain direction alternatives are shown above, and the two chain layer alternatives are illustrated below. In the case of a triple-chain-length structure, there is a possibility of segregation of different types of chains into different layers. In order to achieve this, the corresponding chain direction alternative (shown above) is selected. The triple-chain-length alternative follows the structure proposed for SOS (Larsson, 1972).

obvious relation between chain segregation in different layers and the tendency to form a glycerol group structure with chains in the 2-direction pointing opposite to the other two or the alternative (1- or 3-position together with the 2-position directed in the same way).

Let us consider an example. A triglyceride with a short chain like butyric acid in the 1-position and long chains like palmitic acid in the 2- and 3-positions will probably form a bilayer α-phase. On transitions into β′-and β-forms, the molecules will certainly adopt the triple-chain-length structure, which means that the chains in the 2- and 3-positions pack in one layer and the short butyric acid chains in a separate layer. The conformation of the glycerol group therefore has to adopt the corresponding alternative. The molecular conformation alternative in Figure 2.24 is obvious as is the layer arrangement, *i.e.* the triple-chain-length structure.

Furthermore, we can predict the relative stability of the different polymorphic structures by considering the appearance of the methyl end-group planes. In simple lipid structures with only one chain in the molecules, these end-group regions are always planar, whereas in the case of triglycerides the methyl end groups form some kind of terrace-shaped surface. These terraces must fit into one another without too much empty space. An analysis of the packing constraints gave certain restrictions on stable crystal forms with a double-chain-length structure and a glycerol conformation like that in the β-form of monoacid triglycerides (Larsson, 1966b). This analysis was extended further by de Jong, giving useful relationships on stability conditions of β-forms (de Jong, 1980).

An illustration of how complex the phase properties can be in cases of mixed triglycerides is found in a report on 1,2-dipalmitoyl-3-decanoyl-*sn*-glycerol (Kodali *et al.*, 1989). The bilayer type of structure (double-chain length) is formed when the melt is cooled and the α-form crystallizes. Then two forms which are both β'-polymorphs are obtained on heating. On further heating, melting and solidification into a triple-chain-length β-form takes place (a requirement is a long period of incubation after melting). If the α-form is heated rapidly to avoid β'_2 and β'_3 formation, melting and solidification into a β'_1-form takes place. This form is most remarkable in the aspect that it appears to form a hexalayered structure (long-spacing 111 Å). The authors propose that this structure has a triple-chain-length arrangement of tilted molecules, with alternation of tilt directions in adjacent layers.

Sato and coworkers have studied the polymorphism of StOSt, POP and POSt (O = oleate) in detail (Sato *et al.*, 1989; Arishima *et al.*, 1991). Six polymorphic forms were observed in POP and five in StOSt, the most stable being β_1. The triple-chain-length structure of this form is due to segregation of unsaturated chains in one layer and saturated in another (*cf.* Larsson, 1972), and the proposed arrangement is illustrated in Figure 2.24. The α-form on the other side has a double-chain-structure, which means that the *cis*-double bonds have to adopt a mainly linear conformation (Sato *et al.*, 1989).

Racemic POSt was found to have four crystal forms, and the α-form as well as the most stable β-form were isomorphic with α- and β_1-forms of POP and StOSt (Arishima *et al.*, 1991).

4. Plastic Fats - Growth Morphology and Network Formation

When a complex triglyceride mixture in the liquid state is cooled, high-melting molecules will first tend to segregate and form crystals. The behaviour is complex and not easy to predict, because of molecular compound formation and eutectic effects, but under all conditions crystals will be in equilibrium with the surrounding oil. When the crystalline proportion exceeds about ten percent, a semi-solid material, homogeneous when viewed by the naked eye, is obtained. The oil is thus immobilized by the crystal matrix, a situation analogous to a protein or polysaccharide gel holding water in a network of polymer molecules. Such a solid/liquid mixture is called a *plastic fat*; it shows a mainly viscous deformation with only a minor elastic behaviour on flow. In most application of fats and oils, we are dealing with solid/liquid mixtures - with plastic fats in other words - and it is therefore important to understand their colloidal structure and physical properties.

When triglyceride molecules are transformed from the disordered melt into the solid state, they are first organized in an α-form lattice. This bilayer structure will exhibit a disordered surface structure resembling that of a liquid-crystalline phase (*cf.* the L_α-phase described in Chapter 3). Once a bilayer region of the α-form has crystallized, the molecules in the liquid will "see" the surface of this bilayer, and this surface structure does not reflect the sizes of the triglyceride molecules in the bilayer. Epitaxial growth (which means that a plane of the crystal has identical or very similar periodicity to a corresponding plane in another crystal form) of triglyceride molecules with different acyl chains is therefore possible. This can also be observed in X-ray diffraction patterns from α-forms of complex mixtures; the long spacing lines reflecting each bilayer thickness can cover a range of spacings. When, however, phase transitions into β' - and β-forms occur, the more ordered lattices do not allow the same degree of co-crystallization.

The kinetics of triglyceride crystallization have been studied extensively by Walstra (1987). The most striking feature is the extremely slow crystal growth process. By sufficient supercooling, we can always obtain nuclei immediately, most being formed by heterogeneous and secondary nucleation. The presence of a certain amount of free fatty acids and monoglycerides in a triglyceride oil results directly in surfaces for heterogeneous nucleation, due to self-assembly of the polar lipids (see Chapter 3). The growth process of milk fat under isothermal conditions has been found to take up to 24 hours to be complete (Walstra, 1987). At 11°C, only about half of the final crystallization was reached after one hour. This means that during technical processing of fats, crystallization equilibria will not be reached. An additional factor is so-called Ostwald ripening, which means that there is a molecular rearrangement in crystals of the same composition (and structure), due to growth of the larger crystals as the smaller ones disappear. The reason is the reduction in surface energy.

From this discussion it is clear that nucleation and growth occur in different time scales, and the thermal processing of a fat is an intricate problem if we want to have a certain size distribution of crystals. In food applications, fat crystals, which are platelet-like, must be below about 30μm in the largest dimension to avoid a "sandy' mouth-feeling. A challenging problem is to control crystallization, so that a minimum of solid content can give a homogeneous plastic fat. This is one way of reducing the hydrogenated fat content. Three "new" ways to achieve this seem possible:

(*i*). *The introduction of diglycerides as solid phase.* From a nutritional point of view there can hardly be any objection to a "diglyceride-fat" as it will be digested and absorbed in the same manner as triglycerides. The crystal morphology, however, is quite different. Diglycerides can form thin platelets parallel to the bilayer like all simple lipids, contrary to triglycerides, which are more alike in all dimensions, probably because of the rippled methyl end group structure. One can in fact form a homogeneous fat from a triglyceride oil with only a few percent of diglyceride crystals present.

(*ii*). *Crystallization in supercritical carbon dioxide.* A limitation in the crystallization process is the slow heat transfer on an industrial scale. If the triglyceride crystals are solubilised by carbon dioxide in the supercritical state (above about 73 bar and 31°C), an isothermal crystallization process can be performed through the whole bulk of triglycerides at the same time by reducing the pressure. Such a technique should offer unique possibilities to monitor nucleation and growth, although there are certain complications, *e.g.* removal of heat of crystallization.

(*iii*). *Crystallization at high pressures.* Pressure in itself has a significant effect on triglyceride crystallization, and high pressure effects in food processing are a field which has been studied extensively during recent years. One illustrative example is a report concerning milk fat (Buchheim and Abou El-Nour, 1992). By applying up to 500 MPa of hydrostatic pressure, they followed pressure-induced crystallization of milk fat. An increase in melting point of about 2° per 100 MPa was observed (as also reported in other lipids). Rapid crystallization within the fat globules was thus obtained, a process which normally takes an extremely long time, because of the small size of the emulsion droplets. It might also be possible that crystallization of lipids in the membrane surrounding the fat globule (*cf.* Chapter 14) induce fat crystallization from the fat globule surface and inwards.

The combination of pressure-induced crystallization and the presence of supercritical carbon dioxide should provide enormous variability in crystallization conditions. The main advantage in relation to a thermal process is the velocity; we can at once get the same thermodynamic conditions in a large reactor without being limited by heat transfer. Furthermore there are interesting side effects; the supercritical carbon dioxide dissolved in emulsion droplets can induce fragmentation if the pressure is reduced quickly enough. Pressure variations as a complement to tempering are therefore a type of process which might be expected to be introduced in the future. The generation of heat of crystallization in pressure-induced crystallization (or in pressure reduction when supercritical CO_2 is used) provides obvious limitations in the kinetics of such processes. When a polymorphic transition is obtained in this way, there is a significant increase in melting point. A certain increase in temperature can therefore be permitted.

The morphology of triglyceride crystals and network formation to give a plastic fat is the final point to be discussed. The α-form forms flexible flat crystals without any well-defined crystal surfaces beside the (001)-surface parallel to the bilayer plane. This "soft" morphology might have physiological relevance in relation to surface temperature variations. A subcutaneous fat tissue in mammals might solidify, and seeds of oil-rich plants could crystallize on cooling. The disordered liquid-crystalline character of the α-form should not be expected to damage cell membranes, contrary to β′ or β-forms.

The β′-form forms thin needles, with the needle-direction along the *b*-axis (*cf.* Figure 2.22); the molecular organization is simplest in this direction. The β-form grows in platelets parallel to the bilayers, but the platelets are considerably thicker than crystals of simple lipids like alcohols, fatty acids, and mono- and diglycerides. As mentioned before, this may be related to the rippled methyl end group surface; the ripples will probably induce nucleation and growth perpendicular to the bilayer.

Because of their shape only the β′-crystals are effective in forming networks, as each needle in a disordered array will touch adjacent needles. Aggregation of the crystals increases with their size due to increasing van der Waals attraction. In the light microscope, large crystals can be observed, and the aggregation has been described as showing four textures (Jewel and Meara, 1970):

- spherulites of radially oriented needles
- spherulites of randomly oriented crystals
- clusters; groups of randomly oriented crystals
- association of clusters and/or bundles

The observation that elasticity increases with storage and decreases by agitation has been taken as an indication of formation of stronger primary bonds and weaker secondary bonds between the crystals (van den Tempel, 1961). It is still not known what kind of forces determines the strength of the network. Sintering of the crystals is one alternative, which means that the crystal lattices are fused. There is one obvious argument against sintering; the crystal symmetry is so low. In order for such joints to form between adjacent crystals, they must in fact have an identical orientation in the β-form, whereas a mirror orientation is also possible in the case of the β′-form. A recent careful study of crystal morphology in margarines and shortenings did not give evidence for sintering, and the authors concluded that only "mechanical intertwining‘ of crystals occurred (Chawla *et al.*, 1990).

From what is presently known, we can summarise the colloidal structure of plastic fats in the following way. The needle-shaped β′-crystals aggregate radially-directed in spherulites or parallel in bundles, and α-and β-crystals form similar aggregates lacking crystal orientation. With time, such a network gives an increase in elasticity, which is probably due to an increased degree of direct contact between crystals, although their lattices are differently oriented. We must expect that the liquid triglyceride bilayer

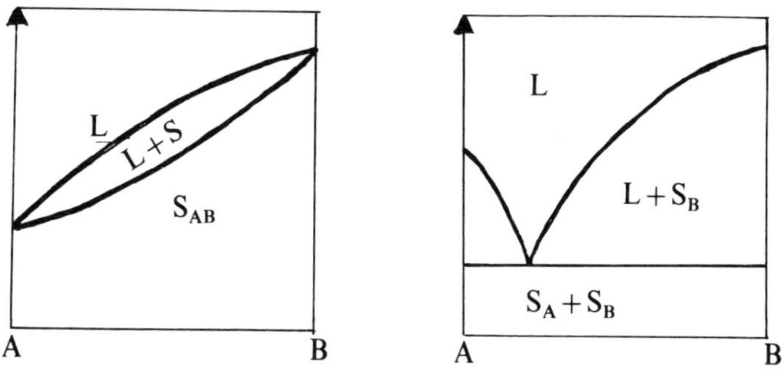

Fig. 2.25. The phase diagrams in the case of complete molecular miscibility (to the left), and no miscibility (to the right). L denotes liquid, S_A:solid of component A, S_B: solid of component B, and S_{AB} : solid when A and B are exchangeable.

(*cf.* Chapter 5), which wets the triglyceride, is rather firmly attached to the surface and gives rise to a repulsive force similar to the "hydration force" (*cf.* Chapter 3). Such a force might in fact be the main factor stabilizing the dispersed state of the crystals. Given time, some crystal growth due to Ostwald ripening will take place, and the increased van der Waals attraction can overcome this liquid-bilayer repulsion to give a stronger association by direct crystal-crystal contacts.

5. *Mixed Triglyceride Systems - Phase Diagrams*

Most studies of crystallization properties of triglyceride mixtures concern binary mixtures, and we will therefore deal mainly with binary triglyceride systems below.

Firstly, phase diagrams, the tool to understand the solid state of mixed systems, will be reviewed briefly. Gibbs phase rule gives the relation between the number of components (C), the number of phases (P) and the degrees of freedom (F) of any mixture.

$$P = C - F + 2$$

If we consider a certain phase, the degrees of freedom are its composition (when it consists of two or more components), the pressure and the temperature. The phase diagram of binary mixtures is drawn like a map, where we can see variations in phases against composition as x-axis and temperature as y-axis. The pressure can also be used as y-axis if the temperature is kept constant. The formula gives the general appearance of a phase diagram.

In binary mixtures, two extreme situations of crystallization behaviour can occur;

- the two components (A and B) can exchange with one another in the crystal lattice at all compositions (mixed crystal formation)
- there are no mutual solubilities in the lattices of crystal A and B respectively.

These two extremes exhibit the phase diagrams shown in Figure 2.25. In

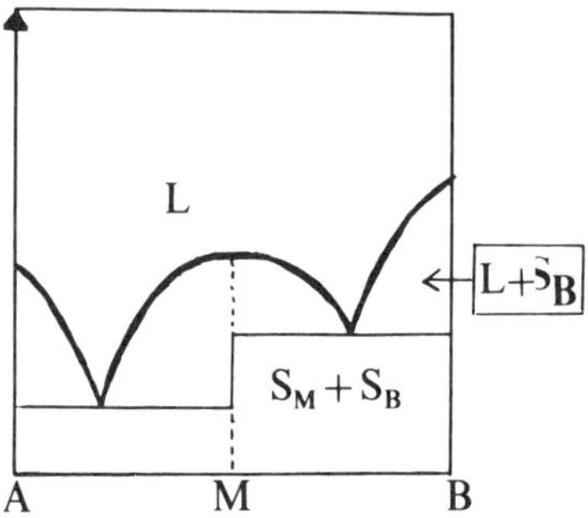

Fig. 2.26. Phase diagram where a molecular compound M is formed of 1:1 composition (the same notations are used as in Figure 2.25).

the case of no mutual solubility in the crystal lattices, the phase diagram exhibits a minimum in melting point at the so-called eutecticum. At the eutectic composition, the liquid crystallizes directly into the two solids.

We also have to consider an alternative type of crystallization. The molecules of the two components A and B can sometimes co-crystallize into a *molecular compound*, which means that at a particular composition, for example 1:1, the two molecules form a stoichiometric complex. With regard to the phase diagram, we then have a situation where the molecular compound, M, can be regarded as a separate component. This is illustrated in Figure 2.26, where a molecular 1:1 compound is formed and there are no mutual solubilities in the crystal lattices. We therefore have eutectic behaviour on both sides, and the phase diagram is in fact the same as if we put two diagrams from Figure 2.25 (the right diagram) together. In this way, any appearance of a binary phase diagram can be traced back and resolved into components of the two diagrams in Figure 2.25.

We are now ready to consider experimental data on triglyceride mixtures. An extensive amount of work on phase diagrams of fatty acids as well as triglycerides was done some years ago by Bailey (1950). Later, Timms (1984) reviewed this field. There is a serious complication in the interpretation of data obtained by just recording melting and solidification points or DSC-curves, and this is caused by polymorphism. An illustration of this is a recent study of the binary system of tristearin and tripalmitin in which the 1:1 mixture was studied in detail (Kellens and Reynaers, 1992). A complete molecular miscibility of the two components in the α-form was observed, and the phase diagram therefore looks like that in Figure 2.25 (to the left). Also, after transition to the β'-form there was complete molecular solubility in the 1:1 ratio. When the β-form is obtained, there is segregation into two forms (tripalmitin + tristearin β-crystals respectively), however. This is seen directly from the X-ray-diffraction pattern (a molecular solution will show a bilayer thickness intermediate to that of tripalmitin and tristearin).

In most triglyceride mixtures of natural origin, which are involved when we deal with oils and fats in foods, we have saturated and unsaturated hydrocarbon chains of different lengths present. Molecular compound formation is then one possibility to achieve effective close-packing arrangements of the molecules in the solid state. An example of this type of behaviour in well-defined model systems is considered below.

Engström (1992) in our laboratory examined the binary systems StOSt/StStO, StOSt/PStO, StOSt/StPO, StOSt/PPO and POP/PPO. Molecular compound formation was observed in two kinds of system: in 2-oleoyl-disaturated/1-oleoyl-disaturated systems and in 2-oleoyl-disaturated/2-saturated-dioleoyl. The 1:1 StOSt/StStO mixture shows a very remarkable melting behaviour (mentioned earlier) on heating from a sub-α form. It gives an α-form which melts at 21°C, and at 32°C the melt crystallizes into a β-form on heating at 20°C/hour. The melted state can last for hours. The proposed structure of this molecular compound is shown in Figure 2.27 together with the structure of the components.

F. GEL PHASES

1. Gel Phases with Crystalline Hydrocarbon Chains

Crystals of polar lipids can form aqueous phases by penetration of water into the polar region of the bilayer, so that the crystals swell without changing the crystalline structure of the bilayer. Such a lamellar structure of alternating lipid bilayers and water layers is illustrated in Figure 2.28. The corresponding phases are called *gel*-phases.

Gel-phases are sometimes thermodynamically stable. One such aqueous system has been characterized in detail; the binary system tetradecylamine-water (Larsson and Al-Mamun, 1974). When water is added to tetradecylamine at 35°C, water just swells the bilayer without any other structural change, as is evident from the X-ray diffraction pattern. The swelling stops when the thickness of the water layer has reached about 14 Å. Above 90% (w/w) of water, however, a second gel phase is formed. The probable reason is that there is then sufficient water to ionize a critical proportion of the amine groups, to give the electrostatic repulsion between the bilayers required for this swelling.

Most polar lipids that form liquid-crystalline phases will also form gel-phases when the temperature is lowered down through the crystallization temperature of the hydrocarbon chain (*cf.* Chapter 3, Section A). One example is monoglycerides. When the liquid-crystalline phase of lamellar monostearin is cooled, a gel-phase is formed when the temperature approaches 50°C. The bilayer thickness of this phase has a thickness of 50.5 Å, and it can accommodate a water layer thickness of 14-15 Å. X-ray diffraction shows the α-form chain packing. If two per cent of a charged lipid like sodium stearate is present, the swelling will proceed, however. In the region at and above 90% (w/w) of water, we can therefore observe the

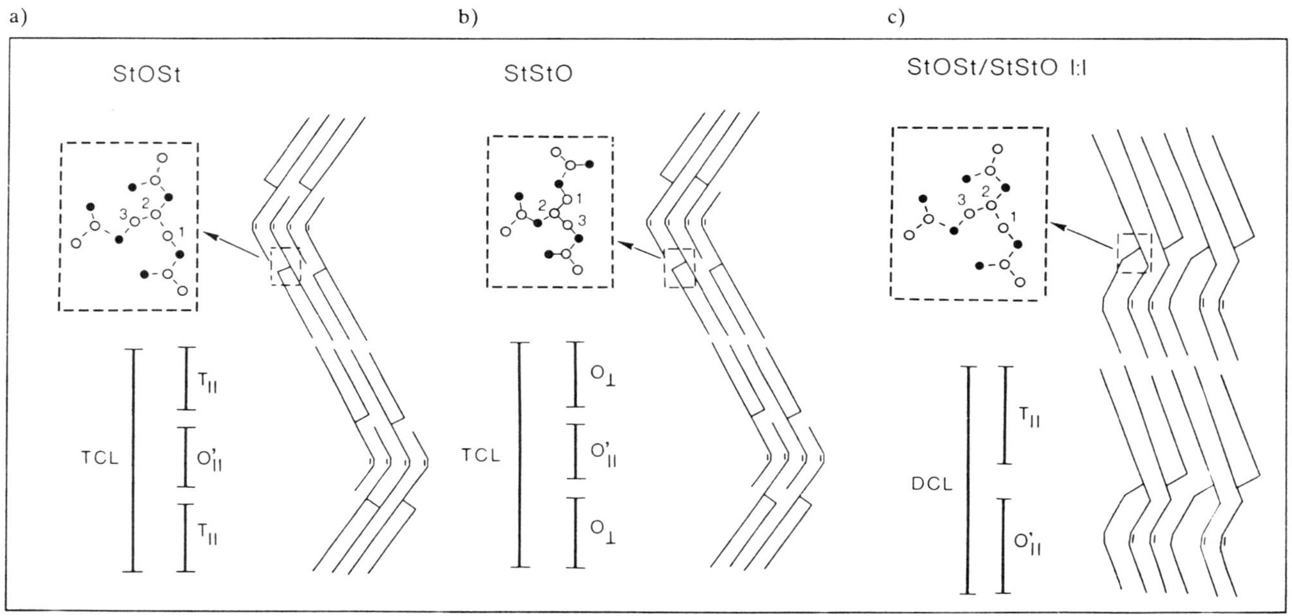

Fig. 2.27. Proposed structure of the stable forms of StOSt under a) (*cf.* Larsson, 1972), of StStO under b), and of StOSt/StStO 1:1 under c), after Engström (1992).

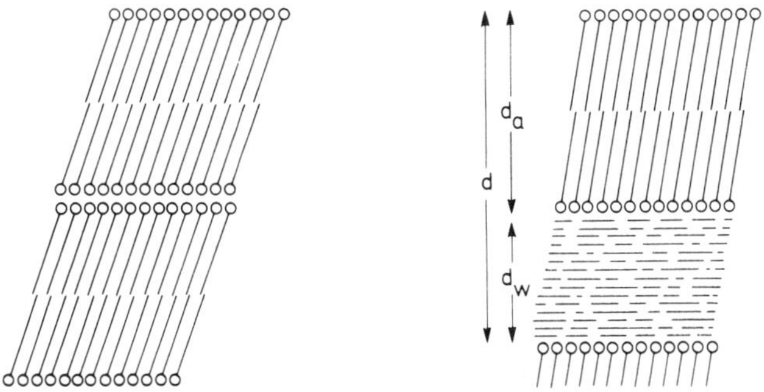

Fig. 2.28. Formation of the gel-state when water penetrates the gap between adjacent sheets of polar head groups of a polar lipid crystal structure. The polar head is indicated by circles. and the chain axis of the molecule is also given.

colours of the visible spectrum when the unit thickness passes the wavelengths of light (Larsson and Krog, 1973). This gel phase is metastable, but it can take up to a year until it is transformed into β-crystals of monostearin + water ("coagel"). The transition is delayed by the presence of impurities solubilized into the α-type of bilayer.

A detailed study of the gel- and coagel states has been reported in aqueous mixtures of dioctadecyldimethylammonium chloride (Laughlin *et al.*, 1990). Below the chain melting point and with increasing amount of water, the anhydrous crystal goes over to a monohydrate and then to a dihydrate. There is therefore no "free" water layer corresponding to a gel-state in this system as was assumed earlier.

2. *Gel Phases with Rigid Hydrocarbon Regions*

Cholesterol phosphate and sulphate occur in cell membranes; cholesterol sulphate is also an important skin lipid component (Chapter 10). The aqueous systems of these lipids have been examined, and they were found to give gel-phases as the predominant aqueous phases (Abrahamsson *et al.*, 1977). Cholesterol sulphate crystallizes as its sodium salt with two molecules of water, and the formation of the gel-phase takes place without any change in the bilayer structure; water just separates the bilayers at the gap between the head groups as indicated in Figure 2.29 (Abrahamsson *et al.*, 1977).

Cholesterol sulphate is not stable for sufficient time in aqueous samples to reach equilibration and allow X-ray analysis, and cholesterol dihydrogen phosphate was therefore studied. It forms a gel phase at room temperature when the water content is above about 84% (w/w) of water, and the reason is probably the same as in the case of tetradecylamine discussed above; water is needed for a certain degree of ionization, required in turn for the electrostatic repulsion and swelling. The swelling

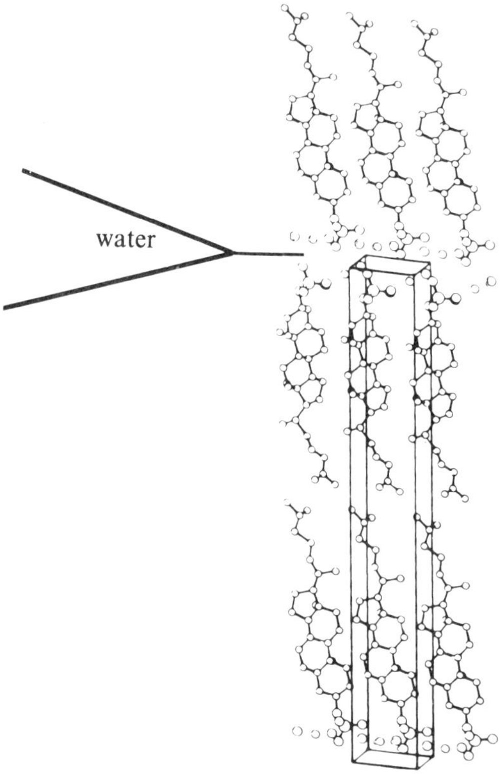

Fig. 2.29. Indication of gel-phase formation when water enters the gap between the polar head groups of crystals of sodium cholesterol sulphate dihydrate, after Abrahamsson *et al.* (1977).

behaviour was followed by X-ray difraction, and about 95% (w/w) of water could be accommodated into the gel-phase, giving a unit layer thickness of about 1000 Å (Abrahamsson *et al.*, 1977).

There is also a food emulsifier behaving like tetradecylamine and cholesterol sulphate with regard to gel-phase formation. Acetylated monoglycerides of C_{16}/C_{18} fatty acids, "Cetodan", kindly supplied by Grindsted Products, give an aqueous gel phase with a low water content as the only aqueous lipid-water phase.

3. Phosphatidylcholine Rippled Phase

Saturated diacyl-phosphatidylcholines exhibit a so-called pretransition before the aqueous gel-phase goes over to the L_α-phase (described in the next chapter) with increasing temperature. The occurrence of two sets of spacings was the basis for proposing a rippled bilayer structure (Janiak *et al.*, 1976). In spite of numerous studies applying various physical methods,

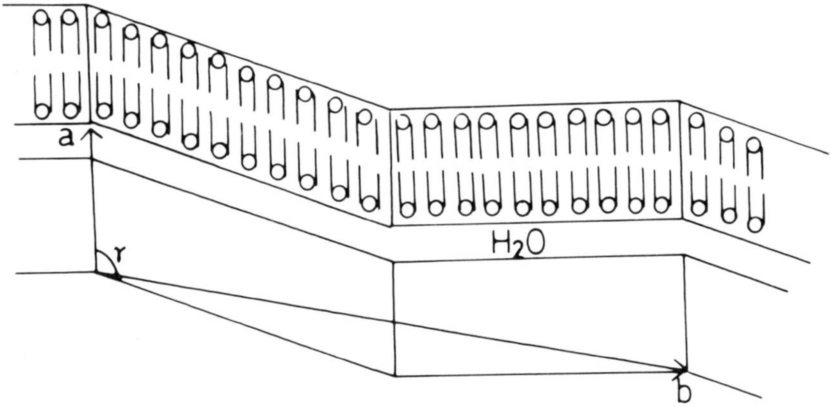

Fig. 2.30. Proposed structure of folded PC bilayers, after Larsson (1977).

the detailed structure is not known. Recent work, has demonstrated the complex kinetics involved in the formation of rippled phases (*cf.* Laggner and Kriechbaum, 1991). Their structure models are based on the assumption of segregation of the molecules into ribbons of two lamellar units, as proposed earlier and shown in Figure 2.30 (Larsson, 1977).

Gel-phases represent an intermediate structure between the solid state and the liquid-crystalline phases considered in the next chapter.

CHAPTER 3

LIQUID-CRYSTALLINE LIPID-WATER PHASES

A. STRUCTURES

1. Introduction

When a 2-monoglyceride of medium chain length is heated, it does not pass directly from the solid state to the liquid state. Just below the melting point, a semi-liquid phase is observed that gives sharp X-ray diffraction lines just like a true crystal. Because of these properties, characteristic for the liquid as well as for the crystalline state, the phase is termed a *liquid crystal.* There are different types of liquid-crystalline phases as will be described in this chapter. They are sometimes called *mesomorphic* phases as they exist between solid and liquid phases.

The structural changes when the *thermotropic* (formed by heating) liquid crystalline phase is formed from 2-monoglyceride crystals are indicated in Figure 3.1. Just as in the crystals, the structure consists of infinite lipid bilayers. The difference is that the chains are "melted" in the liquid crystal, and because of this disorder within the bilayer, there is only a perfect crystalline periodicity in one direction, that corresponding to the bilayer thickness. This structure is therefore described as lamellar, and its formation can be understood in the following way.

The solid state of a polar lipid is characterized by strong forces in the polar head sheets, whereas the forces between the hydrocarbon chains are relatively weak. When such a structure is heated, the van der Waals interaction between close-packed chains ultimately becomes too weak compared to the thermal motions; the chains are transformed into a state of chaotic disorder with a high degree of *gauche*-conformations. The forces between the polar heads are still strong enough to hold them together, however. Thus, the infinite bilayer structure remain unchanged, although the chains have adopted a disordered liquid-like state.

There is another way by which lipids can form liquid crystals. Provided that the temperature is above that where gel phases are formed (see Chapter 2, Section F), the addition of water will result in the formation of a liquid crystalline phase, as illustrated in Figure 3.2. Such liquid

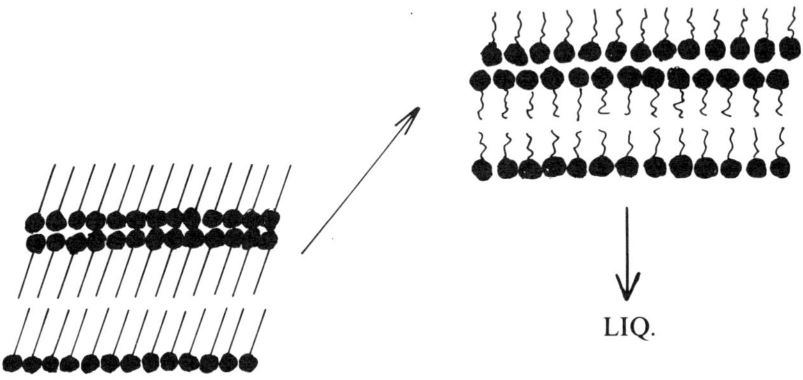

Fig. 3.1. Formation of a lamellar liquid crystal of a 2-monoglyceride just before melting. Molecular polar heads and chains are given as in Fig. 2.28.

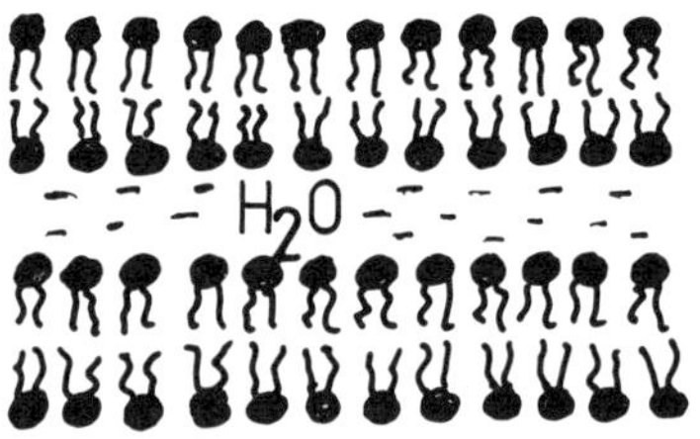

Fig. 3.2. Formation of a lamellar liquid-crystalline phase from crystals of a polar lipid containing two chains by the addition of water (for example, from dimyristoylphosphatidylcholine above 23°C).

crystalline phases are called *lyotropic*, as they are formed when a solvent is added (other polar solvents like glycerol can give the same effect). It should also be stressed that the gel-phase (Chapter 2, Section F) will be transformed into the same phase just by heating, and in the opposite way, cooling of lamellar lipid-water phases often gives gel-phases. This lamellar phase is also termed an L_α-phase. The L_α-phase requires a cross-sectional area of about 30 Å^2 (28-32 Å^2). In crystals of 1-monoglycerides, the cross-sectional area per molecule is 22.5 Å^2. The volume per CH_2-group in a lipid crystal is 23 Å^3, in the α-form about 26 Å^3, and in liquid-crystalline phases as in the liquid state it is about 30 Å^3.

The transition into an L_α-phase means that the polar head group cannot be extended as an elongation of the chain, as in the crystal (see Figure 2.1). According to an examination of space-filling models, both hydroxyl groups should therefore be expected to form a surface towards the water layer. There should also be space for one or two water molecules to be hydrogen bonded to the two carboxyl oxygens on each molecule, as indicated in Figure 3.3.

The milestone in the understanding of the nature of lipid-water liquid

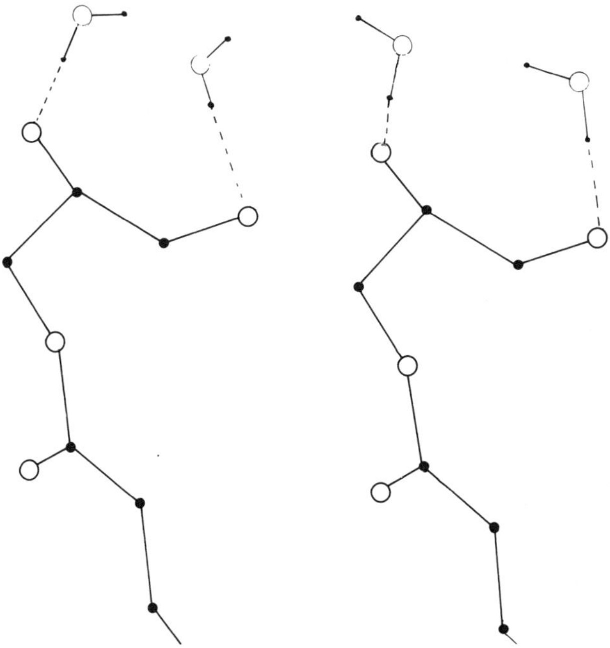

Fig. 3.3. Fragment through the polar region, illustrating a probable head group structure of a monoglyceride L_α-phase, based on the structure in the crystalline state. The tentative positions of the hydrogen-bonded water molecules are indicated.

crystalline phases was the work by Luzzati and coworkers (1960), where the liquid-like character of the hydrocarbon chains was demonstrated unambiguously. It then became obvious that not only lamellar phases can be formed, as shown above, but also the different geometries described below.

2. *Hexagonal Phases*

Luzzati and coworkers (1960) were the first to give a full description of soap-water systems. Soap molecules were earlier known to associate into micelles in water. It was also known that the micelles become elongated at high concentrations. The formation of a hexagonal arrangement of long rod-shaped micelles in water was shown by Luzzati and coworkers to be the explanation for the observed X-ray diffraction lines, corresponding to a two-dimensional hexagonal lattice. The structure derived is shown in Figure 3.4. Very few polar lipids form this structure, but the reversed structure with water cylinders in a lipid matrix is quite common. This structure is also shown in Figure 3.4. They are called H_I and H_{II} respectively.

It is quite obvious what molecular geometry is required in order to form these phases. If the polar head group is dominating in size in

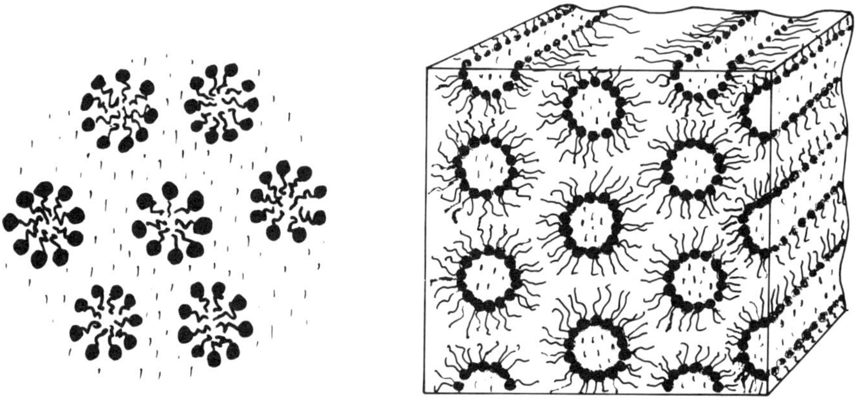

Fig. 3.4. A cross-section of the hexagonal structure H_I (to the left) formed by rod-shaped micellar aggregates of very polar lipids (for example lysophosphatidylcholine) in water, and the reversed structure H_{II}, common among polar lipids (for example, in phosphatidylethanolamine).

relation to the hydrocarbon chain region, the molecules should be expected to form micelles, and then the H_I-phase is observed at increased lipid concentrations. Lysophospholipids form H_I-phases. If on the other hand the hydrocarbon chain region has a large cross-sectional area compared to the polar heads, we should expect that H_{II}-phases are favoured. The detailed shape of the hexagonal H_{II}-phase lipid region is further considered in Section 4 below.

3. *Cubic Phases*

In the pioneering work on liquid-crystalline structures by Luzzati and coworkers (1960), the cubic symmetry in the X-ray diffraction pattern was also identified. The structures of cubic lipid-water phases are quite complex, and the first general model proposed was based on rod systems, with short rod units connected into a continuous network (Luzzati *et al.*, 1968). In strontium myristate, for example, which was analysed in detail, the polar head groups were considered to form the rods in a continuous hydrocarbon chain matrix.

On the other hand, studies of the cubic phase occurring in aqueous monoglycerides indicated a bicontinuous structure, according to NMR-diffusion data, and lipid bilayers as in the L_α-phase, according to geometric relations with the L_α-phase formed at cooling (Lindblom *et al.*, 1979). Our model fitting the probable space group Im3m was a fused system of bilayer units. Later, it was realised that by allowing these units to be curved, the structure of the bilayer becomes an infinite periodical minimal surface (IPMS) (Larsson *et al.*, 1980). Such surfaces possess remarkable properties and can be regarded as a mechanism for uniform packing of lipid bilayers in three dimensions. The IPMS-character of cubic

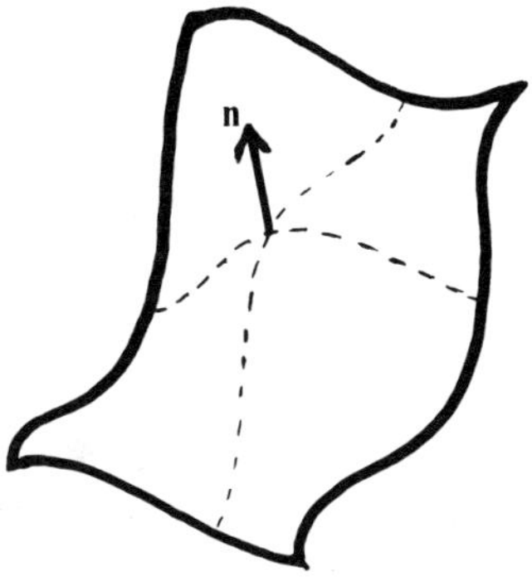

Fig. 3.5. Normal vector (n) to a surface at a particular point.

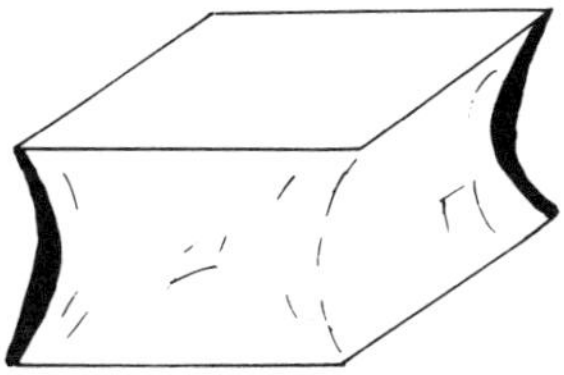

Fig. 3.6. A minimal surface which can be produced by spanning a soap film between two parallel square-frames.

lipid-water phases is now generally accepted, and in order to understand their structure, we will describe their geometry below, starting from curvature.

The curvature of a surface at any point is defined by the principal radii of curvature, R_1 and R_2, which are obtained from the normal vector, n, to the surface, which is shown in Figure 3.5. If we allow the plane through the normal vector to rotate, we can imagine circles which are tangents to our surfaces. The maximum and minimum radius are the principal radii of curvature (R_1 and R_2). The sign defines the side of the surface in relation to the normal vector.

We can then define the average curvature: $H = 1/2R_1 + 1/2R_2$
and the Gaussian curvature K: $K = 1/R_1R_2$

Early mathematics dealt with the curvature of soap films to an appreciable extent, and it was demonstrated that the condition for minimizing a surface spanning a given frame is that the average curvature is zero. The surface in Figure 3.5 is obtained in this way, and it has the minimum area with $H = 0$ in all points. We call such a surface a *minimal surface*. The surface is as convex as it is concave everywhere, and another way to express this is that all points are saddle points. Another example is shown in Figure 3.6.

It is easy to imagine that minimal surface units like that shown in

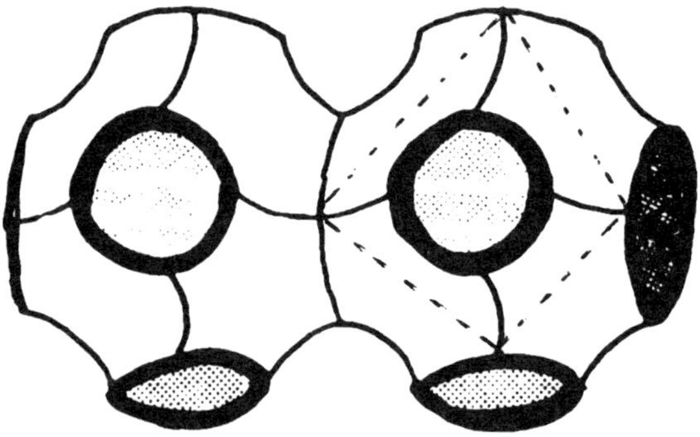

Fig. 3.7. Two structure units of the P-surface fused along the x-axis. The open holes in the y- and z-directions show where adjacent units are located to form a continuous surface. The dotted line indicates one of the square frames which we can imagine in order to form the surface from the units shown in Figure 3.6. The thick lines illustrate the cross-scetions through the continuous surface.

Figure 3.6 can be put together in space to form an *infinite* and *periodic minimal surface*, the IPMS mentioned before. In this way, the simplest cubic lipid-water phase can be formed, as indicated in Figure 3.7. We can assume a system of tubes in the x-, y- and z-directions, which are fused at regular distances along the x-, y- and z-axes. If the geometry at these connections forms a minimal surface, and the volume of the inside of the tube is equivalent to the outside, then a cubic lipid-water structure called a P-surface is obtained. Two such fusion points are indicated in Figure 3.7, and the relationship with the surface in Figure 3.6 is also shown. The mathematician Schwartz derived the P-surface at the end of the last century. The surface just described is the centre of the lipid bilayer, and the tubes are water channel systems on each side of the infinite bilayer. The bilayer has no self-intersection, and the average curvature is zero everywhere. Its symmetry is Im3m (Andersson *et al.*, 1988).

There are three simple cubic IPMS, and all three occur in lipid-water systems. If the tubes are fused alternatively in a tetrahedral geometry, the corresponding IPMS is called a D-surface (corresponding to the diamond lattice). A model of the D-surface is shown in Figure 3.8, and the methyl end group gap of the lipid bilayer is identical to this surface. Again the zero average curvature makes the inside congruent with the outside, and this means that an infinite lipid bilayer separates two equivalent water channel systems. The space group is no. 224 Pn3m (Andersson *et al.*, 1980).

The third surface is the G-surface, where G stands for gyroid. The channel system is a complex helical arrangement with space group no. 230. Ia3d (Andersson *et al.*, 1988), and the structure is shown in Figure 3.9. Considering monoglycerides - our model lipid - they exhibit all three minimal surfaces. 1-Monoolein at a water content of about 20-30% (w/w) gives the G-surface, then the D-surface is found up to a limit of about

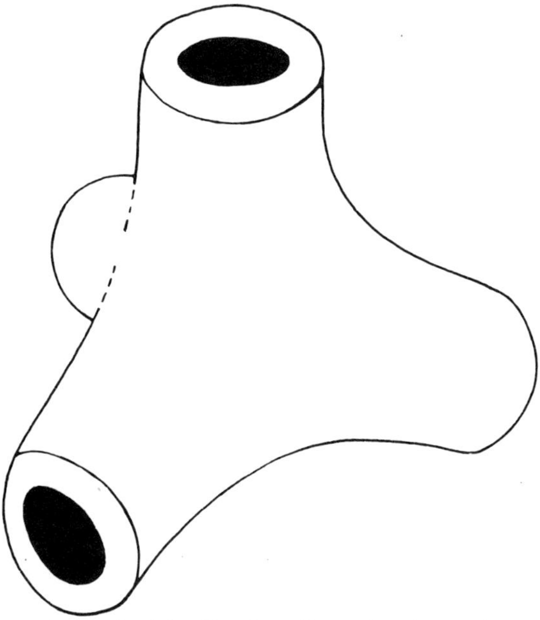

Fig. 3.8. Bilayer structure unit of the D-surface. By putting these units together the whole phase is obtained. The "inside' water is indicated by black.

40% of water (Larsson, 1983). When proteins are present in the water channels, monoolein can also exhibit the P-phase (Larsson, 1986a).

Why are these different cubic phases formed? The average molecular shape in the different phases is considered in the next section. It is obvious that the lipid molecular cross-section parallel to the bilayer is constant in the L_α-phase, whereas it increases towards the methyl end group, due to the chain divergence of the chains, in the reversed hexagonal phase. The cubic phases have molecular shapes in between these extremes, as any surface parallel to an IPMS has a smaller area. This means that the area at the methyl end groups is always larger than that at the polar heads. Furthermore the calculated shape corresponding to a certain bilayer thickness is different in the three alternatives, P, D and G, given a particular lipid-water ratio.

There is another significant factor behind the occurrence of different cubic phases, namely the space requirement of the water channels (Larsson, 1989). If we consider a lipid bilayer centred on the P-surface shown in Figure 3.7, it is obvious that the circular necks between each unit must be wide enough to give space for a lipid bilayer and water. This gives a minimum of water swelling. The limit is shifted to allow a lower water content in the C_D-structure. The C_G-surface is ideal in this respect, as there are no necks of this kind, which means that the C_G-structure may exist down towards zero content of water. There is furthermore an interesting possibility of transitions between these three phases keeping the

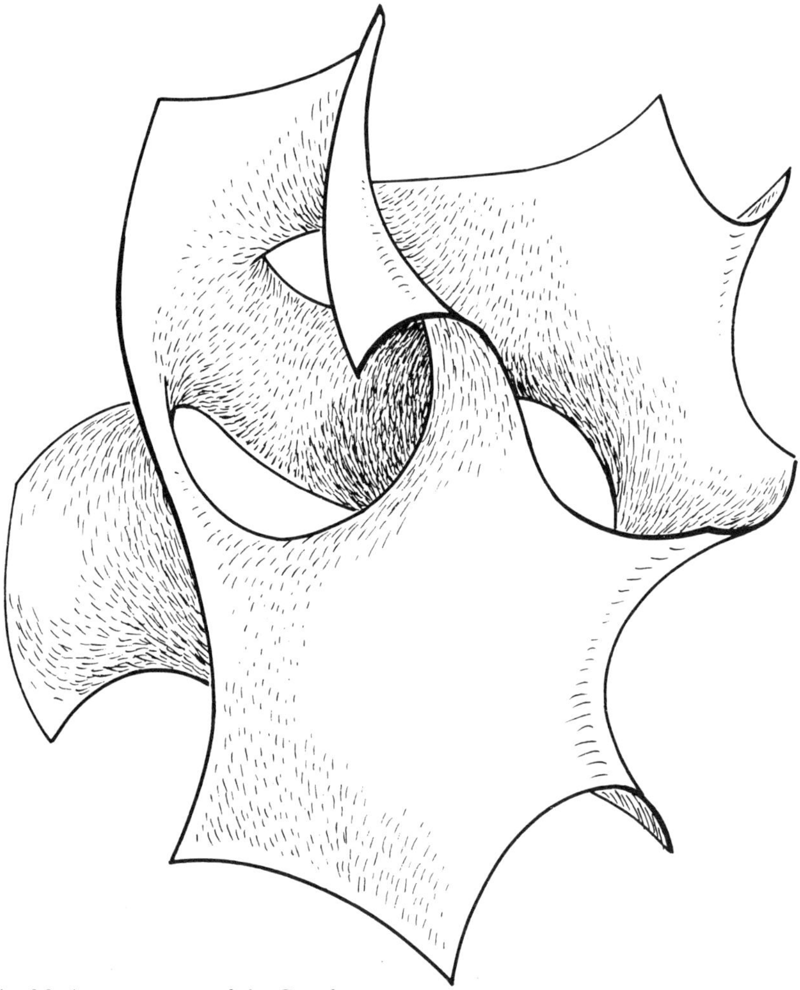

Fig. 3.9. A contour map of the G-surface.

Gaussian curvature (and the average curvature) constant, the so-called Bonnet transformation discussed below.

The minimal surface structure of cubic lipid-water phases possesses many remarkable properties. Perhaps the most important one is the hydrostatic pressure. According to the Laplace-Young equation, there is always a pressure gradient, Δp, over an interface:

$$\Delta p = \gamma \ (1/R_1 + 1/R_2)$$

where γ is the interfacial tension. In a minimal surface, $(1/R_1 + 1/R_2) = 0$ and thus, $\Delta p = 0$. Some new and important biological aspects of these phases will be discussed in Chapter 9. Pressure gradients may then occur, and this means that the average curvature is constant but not zero. The relationship of elastic energy to curvature is discussed below in Section B.

Furthermore two other types of cubic lipid-water phases, which are far less common, will be described. Very polar lipids, such as lysophospholipids, form cubic phases, where the lipid molecules are associated into micelles, and these micelles surrounded by water form a cubic lattice (Arvidsson *et al.*, 1985). The space group is no. 223 Pm3n, and there are two types of micelles, one quasi-spherical and one disc-shaped, according to a recent structural analysis (Varga *et al.*, 1992). All lipids which form this type of structure are water-soluble in a micellar state. Soaps can also form such cubic phases.

An additional type of cubic phase has been characterized by Luzzati and coworkers (Mariani *et al.*, 1988). It is obtained in mixtures of lipids which form aqueous phases with water (without being water-soluble) with lipids that form no such phases, *i.e.* non-polar lipids in our nomenclature. This cubic structure resembles the polyhedral arrangement in a foam with the lipid bilayer corresponding to the foam lamellae. The space group is Fd3m, and the structure is proposed to consist of water micelles in a tetrahedral arrangement.

Finally, it could be added that there have been reports of asymmetric bilayers of IPMS-structures, and on the existence of alternative IPMS-structures in certain lipids. Such alternative IPMS-structures might have the centre of the aqueous layer at the IPMS, so that the lipid molecules form rod-shaped micelles in the channel systems.

B. PHASE TRANSITIONS AND PHASE DIAGRAMS

1. Introduction

The phase rule and the general features of binary systems were discussed in connection with the solid state behaviour of lipids (Chapter 2, Section F). The phase rule applied to aqueous systems exhibiting aqueous phases will be considered here. The functional significance of knowing a phase diagram of a particular lipid-water system is quite obvious. We can then see directly at what water composition and temperature range a certain lipid-water phase exists.

There are different types of phase diagrams, as will be seen below, and after describing the phase properties of various lipids in Section C, we will end this chapter by discussing these different types.

The self-assembly of lipid molecules into micelles was mentioned earlier in Chapter 1. Very few lipids have such a dominating polar region that they are water-soluble. Then, above the *critical micellar concentration* (CMC), the molecules associate into micellar aggregates. We will mainly deal with water-insoluble lipids here, and they can only be *dispersed* in excess of water. Such dispersions are described in Chapter 7.

2. *Why are Aqueous Lipid-Water Phases Formed - The Hydration Force*

In a naive way, we can consider the formation of liquid-crystalline lipid-water phases as a solubility compromise. The polar head groups, which by themselves are water-soluble, are surrounded by water, whereas the insoluble hydrocarbon chains are arranged so that water contact is avoided. With this in view, the water affinity of the polar head groups is the driving force behind formation of the liquid-crystalline phases. The term hydration force was coined by Rand and Parsegian with coworkers on the basis of measurements of osmotic stress on the L_α-phase of phospholipids (de Neveu *et al.*, 1976). A repulsion force versus distance was thus recorded. The curve showed an exponential fall-off with a decay length corresponding to the thickness of a single layer of water molecules. This force was found to be the predominant one in comparison to others (see the DLVO-theory in Chapter 8) at distances below 20 Å.

The hydration force has been the subject of controversy for some time. Recently, Israelachvili and Wennerström (1990) proposed that the force originated from the movement of the lipid molecules in and out from the bilayer plane. Finger like protrusions with a density decreasing exponentially from the polar head group surface would therefore be the reason for the observed exponential repulsive force. Parsegian and Rand (1991) later discussed arguments against this model, and at present there is no generally accepted theory behind the swelling forces which are manifested as formation of liquid-crystalline phases.

3. *Relations between Molecular Geometry and Structure of Lipid-Water Phases*

As mentioned earlier, it seems rather obvious that lipid molecules with large polar heads compared to their hydrocarbon chain region tend to form the H_I-phase, whereas those with small polar heads and a bulky hydrocarbon chain region tend to form the H_{II}-phase. Such simple geometry was used early on in order to explain phase properties (*cf.* Larsson and Lundström, 1976).

With this reasoning, it was easy to understand that the H_{II}-phase could be formed when the L_α-phase is heated, as the thermal movement increases towards the chain end (Larsson and Lundström, 1976). A more sophisticated approach was introduced by Israelachvili and coworkers (1977). They defined a shape factor, v/aL, where v is the molecular volume, a is the cross-sectional area at the water contact region and L is the molecular length. When the shape factor is 1, the L_α-phase is formed, whereas values $\rangle 1$ or $\langle 1$ indicate a tendency to form the H_I or H_{II}-phase respectively. Later efforts to obtain a more detailed description of these transitions involve curvature energy, the cost in free energy to pack the chains in non-homogeneous regions, hydration repulsion and electrostatic

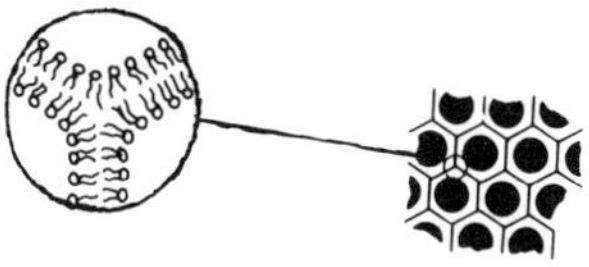

Fig. 3.10. A bilayer description of the H_{II}-phase, contrary to the usual monolayer model (Figure 3.4).

interaction (Kirk *et al.*, 1984). The curvature free energy of a lipid monolayer or bilayer gives the increase in free energy, μ_c, per molecule due to deformation from the spontaneous curvature, R_0, adopted if the layer was free to curl. According to Helfrich (1973):

$$\mu_c = k/2\ (1/R_1 + 1/R_2 - 1/R_0)^2$$

where R_1 and R_2 are the principal radii of curvature and k is the bending modulus of the layer.

The H_{II}-phase has recently been extensively discussed from a theoretical point, in order to understand the common L_α to H_{II} transition in phospholipids (Seddon, 1990). The H_{II}-phase is then described as consisting of two oppositely oriented monolayers. This means that the molecules will build up *frustration* due to the different packing environment of molecules in different locations along the assumed circular cross-section. This is not a unique feature of the H_{II}-phase, however. In the cubic phases, the hydrocarbon packing environment can also vary strongly from flat points with zero Gaussian curvature to points of maximum negative Gaussian curvature, where the chains must be more disordered. Perhaps a better description of the H_{II}-phase is to regard it as a lipid bilayer phase, with intersections of ribbon-like units, as shown in Figure 3.10. The physical/structural techniques available cannot discriminate between these two alternatives. The fact that all the other structural states - L_α-, the cubic, and the L2- and L3- phases and even the liquid state of lipids - are bilayer structures seems to be another good reason to regard the H_{II}-phase as a bilayer structure. On the other hand, the bilayer description is only meaningful when there is enough water present for the polar head groups to form an interface towards water, not towards other head groups.

4. Epitaxial Relations and Phase Transition Mechanisms

The general type of thermal phase transition discussed in connection with solid state polymorphism (Chapter 2) is a first-order transition, where nuclei of the new phase are formed randomly. However, this type of transition seems unlikely in phase transitions involving liquid-crystalline lipid-water phases according to experimental observations. Laggner and Kriechbaum (1991) have recently discussed the transitions as martensitic,

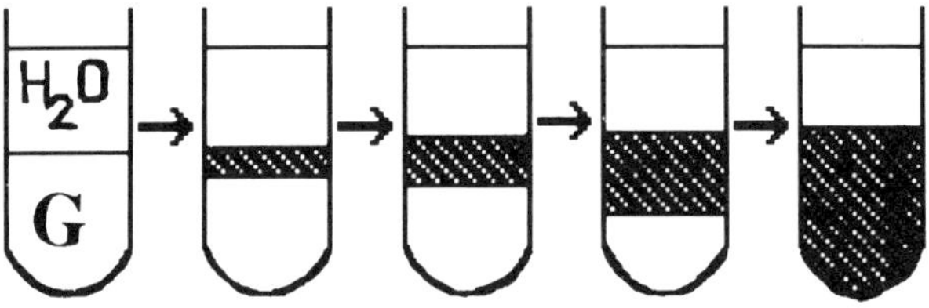

Fig. 3.11 Phase transition behaviour when water is added to the monoolein C_G-phase at its maximum swelling. The shaded region shows formation of the C_D-phase.

a term used to describe iron/steel transitions. This can be clearly observed experimentally. In connection with work on the transition between the two cubic phases in the monoolein-water system (see Section C below), we could see how a sharp front of one phase moved through the other phase without any diffusion zone. A theory on the martensitic transition has also been proposed that involves the so-called Bonnet transformation of the infinite periodic minimal surface structures (Andersson *et al.*, 1986).

The movement of a phase transition front as well as the orientation between the two phases determined by the epitaxial relation is demonstrated in the next paragraph. This direct transition from one lattice to the other means that no surface free energy is involved in forming new nuclei.

5. *Transition between Cubic Phases*

The transition proceeds like a martensitic transition through the bulk, which can be followed easily with monoolein, as shown in Figure 3.11. If we start with the C_G-phase (gyroid-type of structure) of monoolein at its maximum swelling, containing about 35% (w/w) of water, and then add water in excess, a sharp phase boundary between C_G and newly formed C_D can be observed. It reflects some light, although the refractive indices of the two phases are very close. The front moves slowly through the cubic glass-like bulk. At the water front, the C_D-phase consists of about 40% (w/w) of water, and at the C_D/C_P-front it consists of about 35%, so there is a continuous dilatation of the C_D-lattice. X-ray analysis shows that there is an epitaxial relation between the two lattices (Larsson, 1983). Thus a plot of the C_D (111)-line and the C_G(211)-line versus volume-fraction lipid, c/(1-c), forms one line.

A cubic phase always tends to form a single-crystal domain, and when it is exposed to water in excess, a bilayer in the L_α-conformation must be expected to form a surface layer, as this is the only way of avoiding bilayer fractures with the hydrocarbon chains in contact with water. Such a surface layer is shown in Figure 3.12. It is in fact possible to observe interfacial rearrangements towards water, *e.g.* how a rough surface of the cubic phase becomes smooth in contact with excess of water; a process which can be followed over hours in the polarizing microscope.

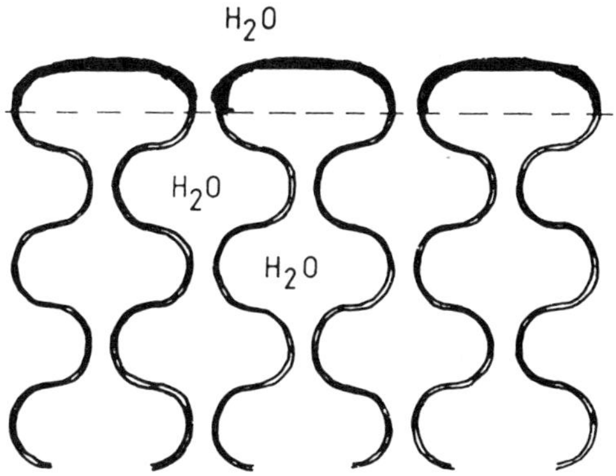

Fig. 3.12. A cross-section through a C_G-phase. The curved lines indicate the centre of the bilayer. The surface towards water with regions of L_α-conformation is indicated by thick lines.

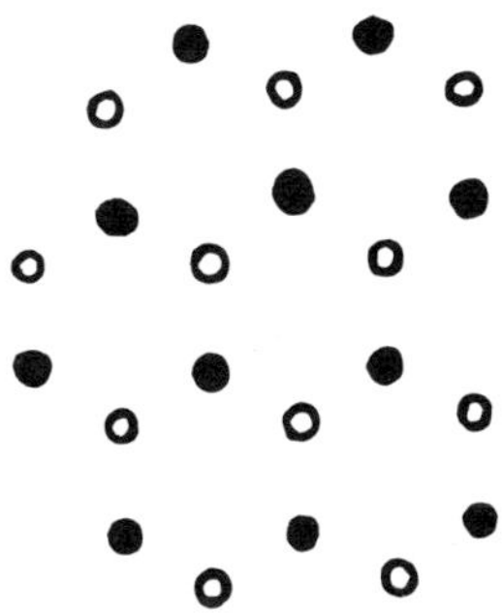

Fig. 3.13. Surface of the C_G-phase as seen from the water side with open circles indicating water channels inwards and filled circles showing where bilayer regions in the L_α-phase conformation plug the channels which point outwards towards the water phase (*cf.* Figure 3.12). The C_D-lattice (111)-plane is identical in geometry with regard to that of channels going up and down. This explains the epitaxial relations seen in a lateral direction in Figure 3.14.

If the surface of the C_G-phase is seen from the water side, it will consist of a lipid bilayer in the L_α-conformation with holes inwards in a hexagonal pattern, as shown in Figure 3.13. When a phase transition $C_G \rightarrow C_D$ then starts at the water contact zone, there is a possibility of epitaxial relations with the C_D-lattice, as indicated in Figure 3.14. The epitaxial growth of the C_D-phase is visualized here by allowing the L_α-interface to invaginate into a C_D-channel system. The C_G-phase is thus "eaten up" by the L_α-front, moving into the C_G-region. The interface between C_G and C_D-regions seen in the test tube corresponds to this hypothetical L_α-layer.

The dilatation of the lattice in the existence range of the C_D-phase is about 10% (from its coexistence with the C_G-phase to its coexistence with excess of water) (Larsson, 1983). Such an expansion can probably be

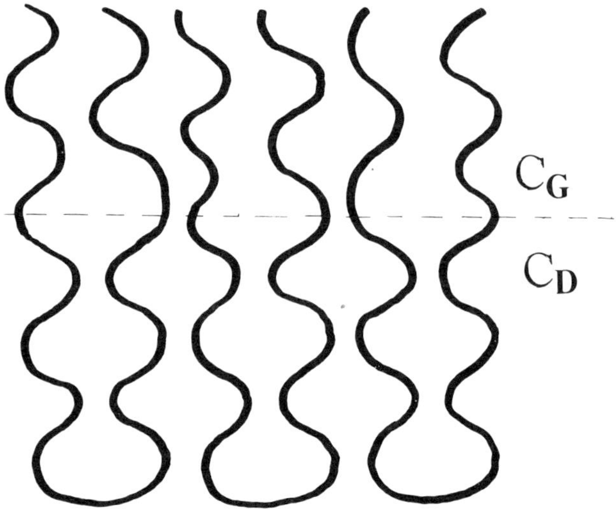

Fig. 3.14. Formation of the C_D-phase at the expense of the C_G-phase. The bilayer is indicated by thick lines.

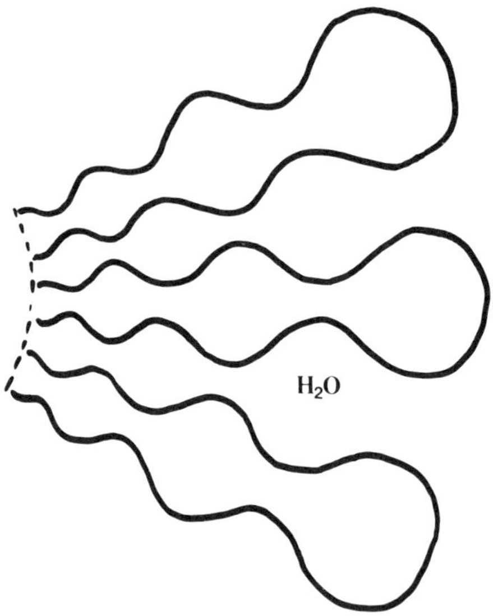

Fig. 3.15. A schematic illustration of dilatation of the C_D-lattice during water swelling. The dotted line shows lattice planes.

accounted for even within a few unit cells because of the deformable character of the bilayer. The dilatation is illustrated in Figure 3.15.

6. *Intermediate Phases*

In surfactant-water systems, the existence of so-called intermediate liquid-crystalline phases has been observed. They occur in the

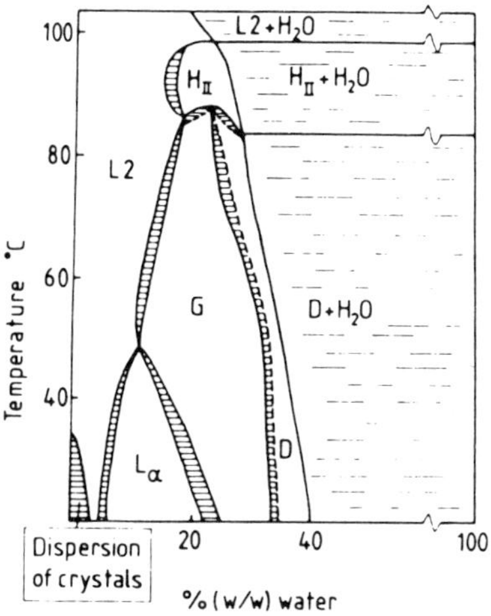

Fig. 3.16. Phase diagram of 1-monoolein-water, after Hyde *et al.* (1984).

concentration range between the L_α- and the H_I-phase. A detailed structural analysis of these phases has been reported in the sodium dodecylsulphate-water system (Kekicheff and Cabane, 1988). It is possible that these phases also exist in aqueous systems of micelle-forming lipids, such as lyso-phospholipids.

C. BEHAVIOUR OF DIFFERENT LIPIDS

1. Monoglycerides

The aqueous system of 1-monoolein is shown in Figure 3.16. The lamellar liquid-crystalline phase L_α is transformed with increasing water content or on heating into cubic phases, and the two cubic phases C_G and C_D occur in a predominant region of the phase diagram. Heating can also give a transition from the cubic to the hexagonal H_{II}-phase. Further heating gives a liquid aqueous phase L2, which will be discussed in the next chapter.

The phase transition sequence obtained by heating illustrates the effect of increased thermal mobility of the hydrocarbon chain; increasing proportions of *gauche*-conformation correspond to increased chain divergence. The shape factor values at the $L_\alpha \rightarrow$ cubic and cubic $\rightarrow H_{II}$ phase transitions can be calculated from the composition and a knowledge of the corresponding X-ray data. The effect of water content on the transitions will be discussed in the last section of this chapter.

A further illustration of the effect of molecular geometry on the phase behaviour is illustrated in Figure 3.17, where saturated 1 monoglycerides from C_8 to C_{20} are compared. It is rather obvious that the effect of chain divergence will increase with chain length; therefore, the H_{II}-phase grows

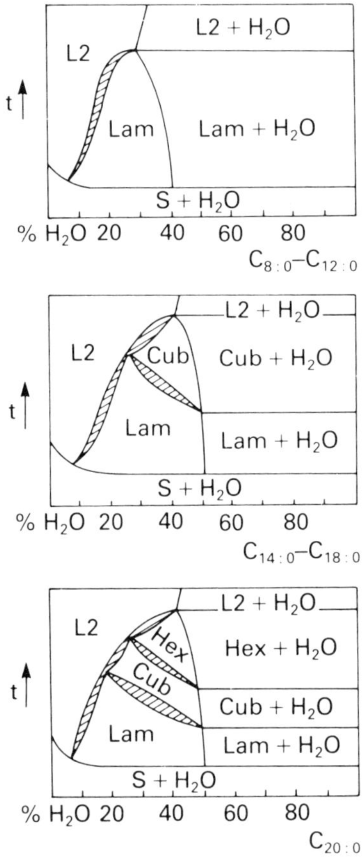

Fig. 3.17. Schematic illustrations of aqueous phase diagrams of saturated monoglycerides from C_8 to C_{20} (Larsson, 1986b).

at the expense of cubic phases, which in turn grow at the expense of the lamellar phase.

In many technical applications, industrially distilled monoglycerides are used. A common type is obtained from fully hydrogenated lard, which has a fatty acid composition dominated by stearic acid. The phase diagram of such a mixture behaves like that of a pure component and is shown in Figure 3.18 (Krog and Larsson, 1968). The transition temperatures are very close to those of 1-monopalmitin (Larsson, 1966). The presence of impurities in the form of charged lipid species results in an increased swelling.

2. *Phospholipids*

Phospholipids, especially phosphatidylcholines, are the polar lipid class which has been the subject of most studies concerning aqueous interaction and phase properties, because of their significance in biomembrane

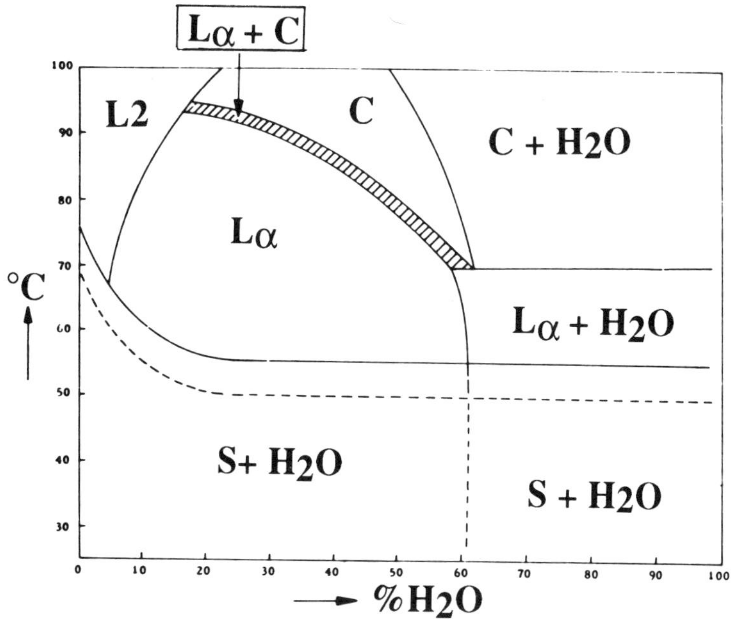

Fig. 3.18. Phase diagram of industrially distilled monoglycerides from fully hydrogenated lard, after Krog and Larsson (1968).

functions. This is illustrated by an extensive literature in the field. A special issue of the journal *Chemistry and Physic of Lipids* was recently devoted to phospholipid phase transitions (Volume 57, 1991).

(*i*). *Phosphatidylcholines* (*PCs*). There is probably no natural lipid that has been studied as thoroughly as egg-yolk PC. The phase diagram is illustrated in Figure 3.19. At a very low water content, the phase behaviour is very complex, and it is omitted here as it is hardly relevant in any application. The lamellar L_α-phase dominates the phase diagram. At low water content, it is possible to obtain the transitions $L_\alpha \rightarrow$ cubic $\rightarrow$ H_{II}, however, as discussed above for monoglycerides. We can explain these transitions in the same way; the driving force is the tendency for increased chain divergence when the thermal mobility increases. Although egg yolk PC is a mixture of several molecular species, it behaves as one component. Other PC mixtures of biological origin, such as PC from soybeans, behave in an almost identical way. Pure synthetic PCs show a similar behaviour.

The phase diagram of dipalmitoyl-PC is shown in Figure 3.20. Apart from a higher chain-melting temperature, the diagram is quite similar to that of egg yolk PC, with a dominating L_α-phase and transitions $L_\alpha \rightarrow$ cubic $\rightarrow$ H_{II} with increasing temperature at a low water content.

When the L_α-phase is cooled, it forms a gel-phase. In a small region of the gel-phase (close to the L_α-transition and at a water content range of about 20-30% (w/w) of water), the so-called rippled phase is obtained. This special case of a gel phase was discussed in Chapter 2, Section G.

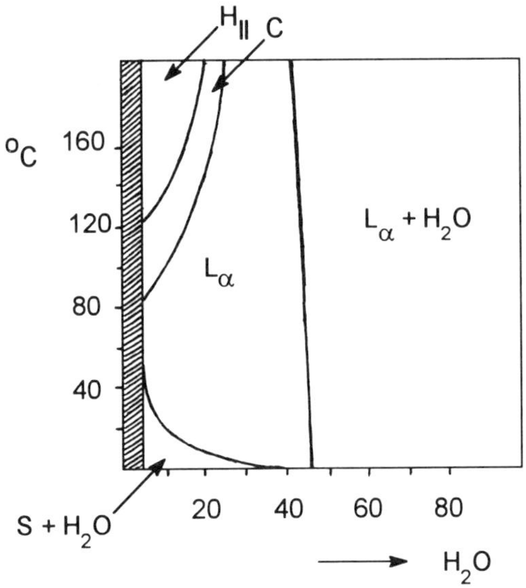

Fig. 3.19. Phase diagram of egg yolk PC, drawn after data from Small (1986).

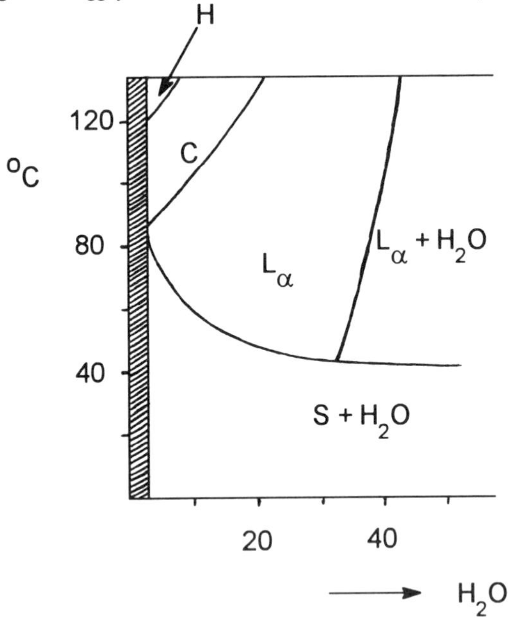

Fig. 3.20. Phase diagram of dipalmitoyl-PC, drawn after data from Cevc (1991).

The transition temperature from the L_α-phase to the gel-phase on cooling goes from 41°C to 23°C when the chain length is reduced from $C_{16:0}$(*cf.* Figure 3.20) to $C_{14:0}$. For $C_{12:0}$ and in the case of $C_{18:1}$ (dioleyl-PC), this transition is below 0°C (Lewis *et al.*, 1987).

A recent study of PC with one $C_{16:0}$ chain and the other short shows

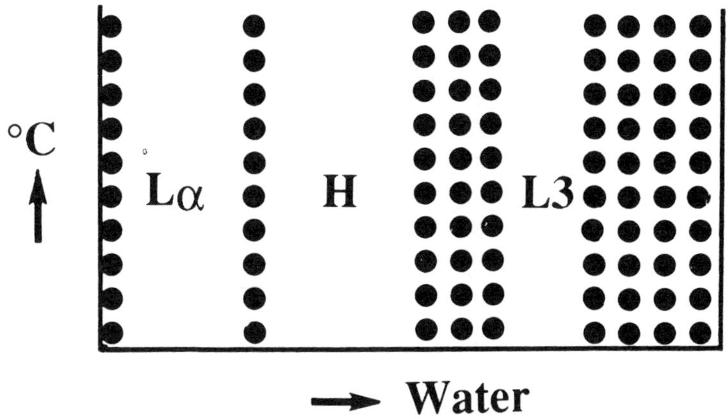

Fig. 3.21. Main features of the phase diagram $C_{6:0}$ $C_{16:0}$ -PC / water. The dotted regions denote two-phase regions, after Svensson *et al.* (1993).

that down to C_8 the phase diagram has the same general appearance with a dominating L_α-phase (Svensson *et al.*, 1993). When the short chain reaches C_6, however, a quite different phase diagram is observed, as shown in Figure 3.21. As we only had a very limited amount of lipid, it was not possible to determine it fully. The most remarkable feature is the occurrence of the hexagonal H_I-phase and an L3-phase, the structure of which is discussed in the next chapter. The phase diagram is in fact more similar to that of lyso-PC, discussed below, than to those of other PCs.

(*ii*). *Phosphatidylethanolamines* (*PEs*). Crystals of didodecyl-PE heated in excess of water are transformed into the L_α-phase $(+H_2O)$ at about 40°C. At about 100°C, a cubic phase $(+H_2O)$ is formed and above 120°C, the hexagonal phase $H_{II}(+H_2O)$ is obtained (Seddon *et al.*, 1982). The binary system dipalmitoyl-PC/water shown in Figure 3.20 exhibits similar transitions. A phase in between the L_α-phase and the cubic phase was not identified, but in consideration of the close agreement with the diagram reported by Seddon and coworkers (1984), it must be a cubic phase. The phase diagram is shown in Figure 3.22.

There are also reports of different PEs with unsaturated chains that exhibit cubic phases and H_{II}-phases (Lindblom and Rilfors, 1989).

When PE/water L_α-phases are cooled to temperatures below the chain crystallization region, there is no stable gel-phase, but a separation into crystals plus water (coagel). This difference in comparison to PCs has been related to the cross-section of the molecules, which reflects the polar head-group packing (Hauser *et al.*, 1981). In PCs, the cross-section is 47-54 Å^2/molecule, which is so close to that of the L_α-phase that the transition is possible without disruption of the head-group lattice. This is not possible in PEs, however, where the cross-section is 35-42 Å^2/molecule.

PEs like PCs are zwitter-ionic, and at high pH they therefore become anionic, whereas they are cationic at low pH. A further discussion of these phenomena is given in connection with mucous surface phospholipids in

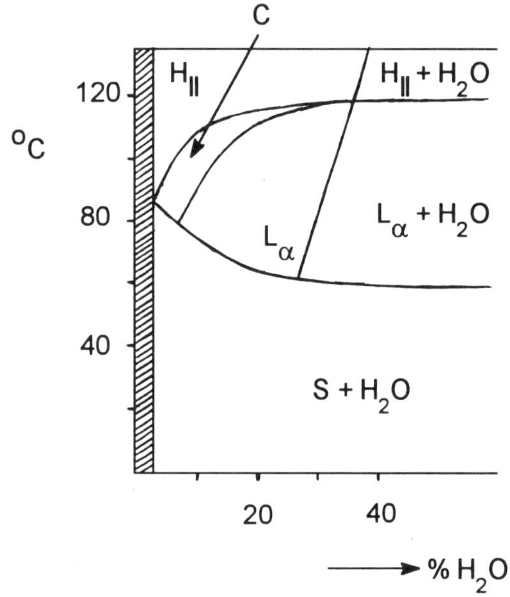

Fig. 3.22. Phase diagram of the binary system dipalmitoyl-PE/water, drawn by combining data from Cevc (1991) and Seddon *et al.* (1984). The complex region of low water content is shaded.

Chapter 12. In the ionic state, these lipids with monovalent counter-ions behave in a similar way to ionic phospholipids discussed below.

(*iii*). *Phosphatidylinositols* (*PIs*). The phase diagram of the aqueous system of soybean PI is shown in Figure 3.23. The hydrocarbon chains are predominantly $C_{16:0}$ and $C_{18:2}$. From room temperature up to 70°C, the L_α-phase dominates the phase diagram (Söderberg, 1990). The limit of swelling is about 77% (w/w) of water.

(*iv*). *Phosphatidylglycerols* (*PGs*), *Phosphatidylserines* (*PSs*) *and Phosphatidic Acid* (*PA*). The behaviour of aqueous mixtures of these anionic phospholipids depends upon the electrostatic repulsive forces, which in turn are related to pH and counter-ions. When these molecules have a negative charge and the counterion is monovalent (the sodium salt is the most common form), the swelling behaviour is similar to that shown in Figure 3.23 also. When the molecules are neutralized, the lateral repulsion between the polar head groups is reduced. They therefore tend to form an H_{II}-phase (according to the change in molecular shape discussed above), as has been shown in the case of PSs (Hope and Cullis, 1980). When divalent ions are present, they tend to contract the polar head groups, giving the same effect. The formation of an H_{II}-phase by the presence of calcium ions has been demonstrated in PGs (Harlos and Eibl, 1980) and in PAs (Farren *et al.*, 1983).

(*v*). *Lysophospholipids.* Type A phospholipases perform a controlled cleavage of phospholipids in membranes to form lysophospholipids and

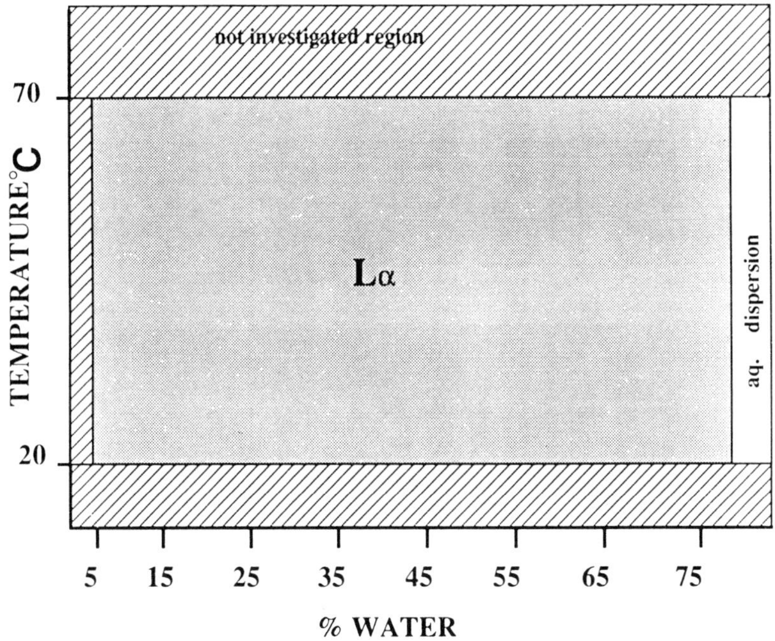

Fig. 3.23. The binary system of the sodium salt of PI-water, after Söderberg (1990).

free fatty acids. This reaction is the basis for physiologically important cascade reactions, and the formation of arachidonic acid (leading to prostaglandins) is the most significant result. An uncontrolled formation of lysophospholipids, induced by snake venom for example, can lead to solubilization of cell membrane lipids and cell lysis. The detergent-like properties of lysophospholipids are responsible for such effects.

Four lyso-PCs have been studied in detail with regard to aqueous phase behaviour (Arvidsson *et al.*, 1985). 1-Palmitoyl-, 1-stearoyl-, 1-oleoyl- and 1-linoleoyl-*sn*-glycerol-3-phosphorylcholine all give a micellar solution, which at increased concentration in water give the hexagonal H_I-phase and the L_α-phase. At intermediate concentrations, cubic phases were also seen. 1-Palmitoyl-lyso-PC shows a cubic phase between the H_I-phase and the micellar solution, whereas the other lyso-PCs shows a cubic phase between the H_I- and the L_α-phase.

It is interesting to note that C_{16}-lyso-PC behaves very similar to the $C_{16}C_6$-PC discussed above. The difference is that the lyso-PC shows a cubic phase, whereas $C_{16}C_6$-PC shows an L3-phase. As will be discussed in the next chapter, the L3-phase is closely related to cubic phases.

3. *Sphingolipids*

Sphingomyelin is rather similar to PC from a chemical point of view (see Chapter 1). The aqueous system is also quite similar with a dominating

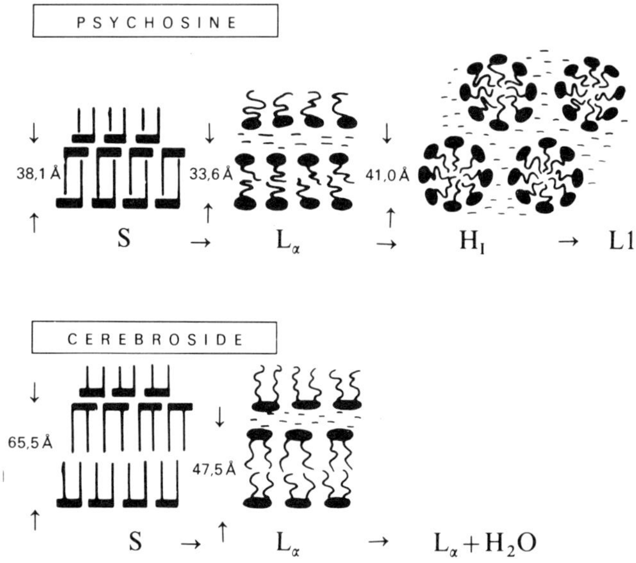

Fig. 3.24. The molecular arrangement of different phases of cerebroside from bovine brain and the corresponding psychosine, after Abrahamsson *et al.* (1972).

L_α-phase showing a limited swelling. Zwitter-ionic lipids in all known phase diagrams give the same swelling limit as non-ionic lipids, with a water layer thickness of about 20 Å. Unsaturated and saturated members have been examined, and they behave in a similar way except for a shift in chain melting point (Shipley *et al.*, 1985; Sripada *et al.*, 1987).

Cerebrosides from bovine brain give an L_α-phase above 70°C, and no other liquid-crystalline phase was observed (Abrahamsson *et al.*, 1972). The corresponding compound obtained when the fatty acid group is removed is called *psychosine*, and it is detergent-like as are lysophospholipids. Thus micellar solutions are formed at high water content, and with increasing concentration the hexagonal phases H_I and the L_α-phase are obtained. The different phases of psychosine and cerebroside are shown in Figure 3.24.

An ionic cerebroside in the form of a sulphatide sodium salt gives a micellar solution at high water content (Abrahamsson *et al.*, 1972), and some gangliosides behave in the same way (Curatolo *et al.*, 1977).

4. Glyceroglycolipids

These lipids are especially abundant in plant cell membranes, and they are often found in microorganisms. Monogalactosyldiacylglycerols (MGDG) form H_{II}-phases with water, whereas digalactosyldiacylglycerols (DGDG) form L_α-phases. The same behaviour was found in MGDG and

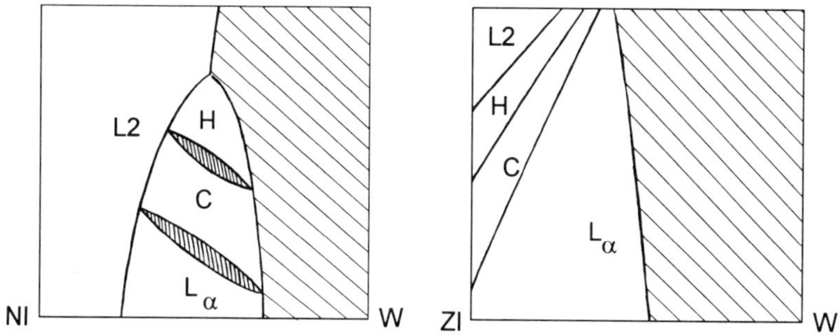

Fig. 3.25. Schematic illustration of one-phase regions in binary aqueous systems of non-ionic lipids (to the left) and ionic or zwitter-ionic ones (to the right). There is a shift in position of the line showing maximum swelling when zwitter-ionic and ionic lipids are compared.

DGDG from plant leaves (Shipley *et al.*, 1973), from wheat endosperm (Larsson and Puang-Ngern, 1979) or from mycoplasma (Wieslander *et al.*, 1978).

The thylakoid membranes of chloroplasts contain MGDG and DGDG as the major components, with some phospholipids. The proportions between the components is close to that corresponding to a transition from L_α- to the H_{II}-phase. In mycoplasma, there is a similar MGDG/DGDG balance close to a phase transition (Wieslander *et al.*, 1978).

D. BINARY AND TERNARY PHASE DIAGRAMS

1. Simplified Model Diagrams

We can separate three different types of phase diagrams of polar lipid-water systems. The strongly polar lipids, such as lysophospholipids, form micellar water solutions (L_1). These systems usually form an H_I-phase and an L_α-phase when the lipid concentration is increased. Cubic phases can also occur between these phases.

The non-ionic polar lipids and the zwitter-ionic or ionic lipids respectively form the other two types of phase diagram, that are indicated in Figure 3.25.

We have illustrated cases when L_α-, H_{II}- and cubic phases all occur in the same systems. There is a remarkable difference in the direction of the phase boundary, and this is the reason why I think that they should be classified as different types. When the lipid is non-ionic, the tilt of the border lines corresponds to a tendency for $L_\alpha \rightarrow$ cubic $\rightarrow H_{II}$ with increasing water content and increasing temperature. We discussed this transition sequence, induced by heating, in connection with monoglyceride-water systems; the tendency towards chain divergence with increased temperature seems rather obvious. The effect of increased water content might be explained in this way. The polar head groups get more freedom to move up and down in relation to the plane of the water layer, and this movement will induce disorder of the chains towards more *gauche*-conformation; thus resulting in increased divergence. The polar head groups are held together laterally by hydrogen bonds so that the cross-section there tends to be constant. Such a mechanism is confirmed by X-ray data; the bilayer thickness decreases with increased water content.

When the polar head groups are kept together by ionic forces, as in the case of ionic (or zwitter-ionic) lipids, there will be a quite different effect when the water content is increased. If a lipid with one charged group only is considered, it is obvious that there is an electrostatic repulsion between molecules arranged laterally in the bilayer. With increasing water content, there will be possibilities to reduce the repulsive forces if the polar heads get more separated. This is now possible as water can penetrate between the polar heads to fill up space. A predominant change with increased water content will thus be an expansion of the molecular cross-sectional area at the polar head groups. This is equivalent to a tendency for change when the water content is increased in the sequence, $H_{II} \rightarrow$ cubic $\rightarrow L_\alpha$. The effect of temperature is the same as in non-ionic lipids. The presence of free counter-ions or polar head groups, which are zwitter-ionic, will not change these arguments. There is one significant difference between zwitter-ionic and non-ionic lipids, however, that is the thickness of the water layer of the L_α-phase. Ionic lipids swell to a very high water content, whereas zwitter-ionic lipids behave like non-ionic ones in this respect, giving a swelling limit at a water layer of about 20 Å. This is related to the geometry of the polar sheet; there is no electrical double-layer, *cf.* Chapter 8 (DLVO-theory).

2. Ternary and Quarternary Phase Diagrams

When a biological or technical application of lipids is considered, there is seldom a situation that involves only binary systems. It has been illustrated above that complex mixtures of one lipid class, such as different PCs, behave like a single component. This simplifies functionality discussions. Many applications involve the presence of different classes of lipids, however. In order to characterize the behaviour of such complex lipid mixtures, it can sometimes be enough to map the phase properties of a ternary mixture: non-polar lipids/water/polar lipids. This approach will be illustrated in the case of cereal lipids, which we will apply in Chapter 14. A useful model system, illustrating microemulsions and considered in the next chapter, is the ternary system monoglycerides (from sunflower oil)/soybean oil/water, which is shown in Figure 3.26.

The incorporation of triglyceride molecules into the L_α-phase will induce transitions into the cubic, then into the H_{II}-phase, and finally into the L2-phase, which will be discussed in the next chapter.

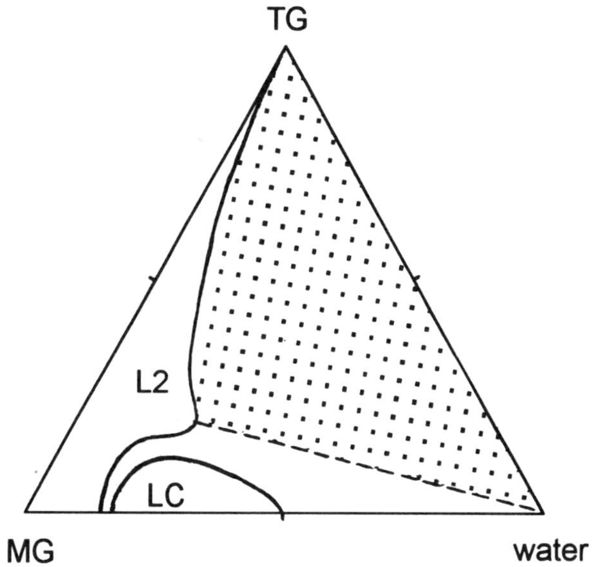

Fig. 3.26. Ternary phase diagram at 40°C of soybean triglycerides (TG)/water/sunflower oil monoglycerides (MG). The liquid-crystalline phases (LC) and the L2-phase are shown. The important two-phase region L2 plus water is dotted, after Lindström *et al.* (1981).

If one component of the polar lipids forms a micellar solution, the phase behaviour should be described by a separate type of phase diagram, with this strongly polar lipid fraction on a separate composition axis. This means that a space representation, using a tetrahedron, is needed to describe phase equilibria in a system consisting of the four components "ordinary" polar lipids/strongly polar lipids/non-polar lipids/water. Let us consider one such case, the quarternary system obtained when saponin is added to the three components of Figure 2.26. Saponins are lipid-like molecules consisting of a steroid skeleton with sugar groups attached. They form micellar solutions and behave in the same manner as bile acid salts from a physical point of view. The ternary system of monoglycerides/saponin/water is shown in Figure 3.27.

The two triangles of the tetrahedron, shown in Figs. 3.26 and 3.27, illustrate the most important features of this four-component system. Thus the ternary diagram for saponin/oil/water shows no liquid-crystalline phase, and solid saponin is present throughout the ternary system saponin/monoglycerides/oil. This example may illustrate how a very complex phase behaviour of a lipid mixture can be resolved by chemical separation into lipid classes (usually three is enough), and then the phase equilibria are mapped in ternary phase diagrams.

Many technical applications involve polar lipids together with *small* amphiphilic molecules in water media. Ethanol, representing a water-

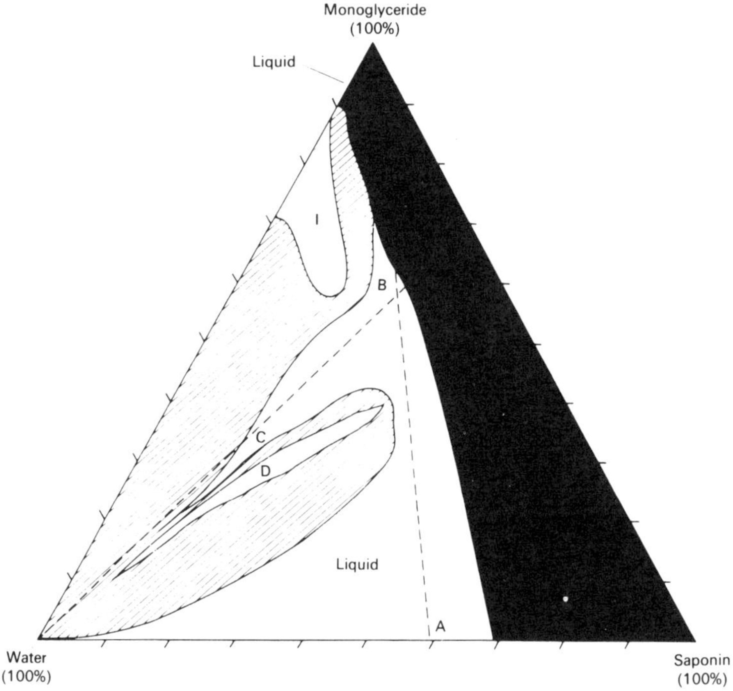

Fig. 3.27. The ternary phase diagram saponin/water/sunflower oil monoglycerides (Barla *et al.*, 1979). The black region shows where saponin crystals are present. The 'I'-region is a cubic phase and the 'D'-region is an L_α-phase.

soluble amphiphile, and *n*-butanol, with limited solubility in water, will be considered below. The ternary system of soybean PC/ethanol/water is shown in Figure 3.28 (Söderberg, 1991). At a high PC content, only a minor proportion of ethanol is needed to transform a liquid-crystalline PC-phase into an L2-phase. Another remarkable feature exhibited by this system is the physical properties of the dispersed L_α-phase. With an increasing ethanol content, the life-time of the dispersed particles (liposomes) is strongly reduced.

The main features of the ternary system of egg yolk PC/*n*-butanol/water is shown in Figure 3.29 (unpublished study from our department). The existence of a single liquid phase from the water to the butanol corner shows that the structure in the liquid state changes successively from that of an L1- to that in an L2-phase. The well known efficiency of so-called water-saturated butanol (WSB) in lipid extraction of biological tissues, *e.g.* cereal grains, can be understood from this dual character of the liquid state. WSB is a 20:80 (v/v) water:*n*-butanol mixture, forming two phases. When some lipid is dissolved in this mixture, it can be seen in the diagram that one phase is formed. The phase diagram also illustrates the effect of solubilization of butanol in the L_α-phase bilayer. At a certain molecular ratio, the curvature is changed and a transition into the H_I-phase takes

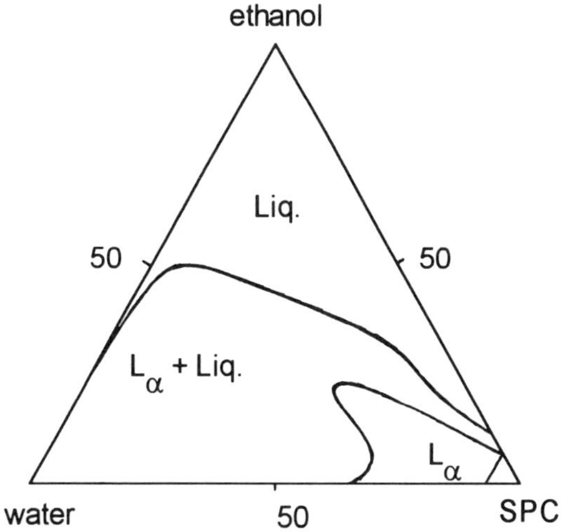

Fig. 3.28. Ternary phase diagram soybean PC/ethanol/water, after Söderberg (1990).

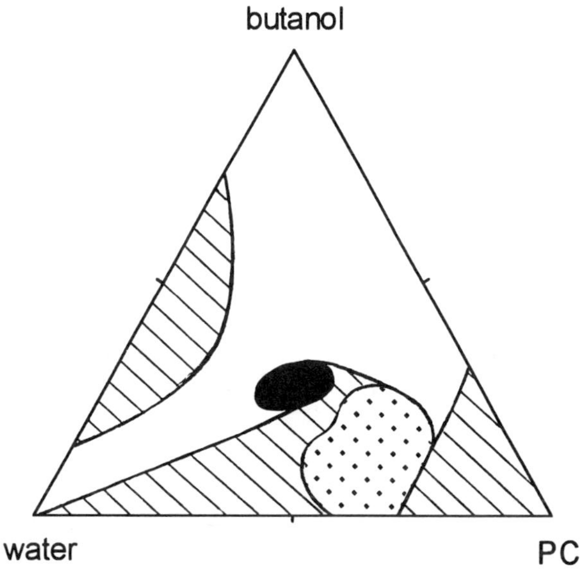

Fig. 3.29. Ternary system egg yolk PC/*n*-butanol/water. The large empty region is one liquid phase, the dotted region is the L2-phase and the black region is the H_I-phase. The shaded regions consist of two phases.

place. The average molecular cross-section at the polar heads increases, and at the hydrocarbon chain ends it decreases, with increasing butanol content.

Mixtures in which butanol was replaced by butyric acid showed similar phase equilibria. This system is of physiological interest as butyric acid is formed by fermentation in the colon and is taken up by the epithelial cells.

E. CHOLESTEROL ESTER LIQUID CRYSTALS

Fatty acid esters of cholesterol are important components of blood lipoproteins (see Chapter 7), and they are able to form unique liquid-crystalline phases. Before melting, they can form smectic, cholesteric and blue phases related to the fatty acid chain length. The smectic phase is considered to be similar to the L_α-phase, whereas the cholesteric phase exhibits a successive twist of the molecules when passing from one layer to another. The blue phases are cubic, and their structures can be described by the IPMS-concept, explained above in connection with the cubic lipid-water phases (*cf.* Andersson *et al.*, 1994). There are three blue phases (BPI, BPII, BPIII) formed successively with increasing temperature. Their structures are closely related as is obvious from the similarities in physical properties and the small transition enthalpies.

These phases are really blue, and the colour reflects a very long periodicity in the molecular organization (the wavelength of blue light is about 4900 Å). The space group has been determined from the crystal surfaces, and BPI has space group I $4_1 32$. There is only one simple IPMS which is chiral, and it has just this space group (A. Fogden, personal communication). One structure model of the blue phases is based on a molecular arrangement in concentric IPMS, with a molecular twist between adjacent layers (Andersson *et al.*, 1994). The layers outwards taken together give the unit cell, with a cell edge corresponding to the wave length of blue light. An alternative structure is just a single IPMS with dimensions of the unit cell corresponding to molecular dimensions. The blue colour is explained if the cholesterol skeletons form helical rods on each site of the minimal surface, and if there is a small twist in the direction of each cholesterol skeleton in relation to its neighbours so that the wavelength of blue light corresponds to a twist of 360°.

CHAPTER 4

THE LIQUID STATE

We have in Chapter 3 described the well-known micellar solution, which is formed in aqueous systems of strongly polar lipids behaving like surfactants. There is also a reversed type of structure, which can be formed by other polar lipid molecules in water, where water aggregates occur in a hydrocarbon chain continuum. To distinguish between these two micellar solutions, they are termed L1- and L2-solutions respectively. The L1-phase is thus connected with the water corner of the phase diagram, whereas the L2-phase is connected with the lipid corner. The occurrence of L2-phases was demonstrated in Chapter 3 in many phase diagrams.

A third liquid phase different from the L1- and L2-phases has recently been characterized in certain lipid-water systems. We observed such a phase with about 70% (w/w) of water in aqueous systems of wheat lipids (see Chapter 14). Lacking a name we termed this phase L2, but now such phases are termed L3-phases. A phase diagram involving an L3-phase was also shown in Chapter 3 (Figure 3.21). The structure and properties of these liquid phases in lipid-water systems will now be described.

The L1-phase, the ordinary micellar solution, has a very limited role in lipid-water systems, and will not therefore be discussed further below. In dilute solutions, the micelles are usually spherical, with a core of hydrocarbon chains disordered as in liquid-crystalline phases. When the concentration of surfactant (or polar lipid) is increased, rod-shaped micelles are usually formed.

Oils are of great technological importance and they will be considered first.

A. OILS - THE LIQUID STATE OF TRIGLYCERIDES

Its is an old experience in the processing of fats and oils that a "memory" of the crystalline state can remain in an oil after melting of fat crystals. Thus keeping the oil for different lengths of time at different temperature above the melting point can result in variations in the polymorphic behaviour. Based on the X-ray scattering in the liquid state of triglycerides, an association of molecules into bilayer regions was

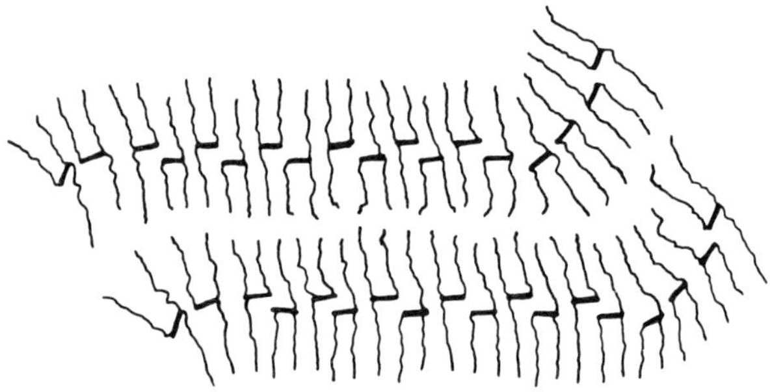

Fig. 4.1. Proposed bilayer arrangement in the liquid state of triglycerides, after Larsson (1972).

proposed (Larsson, 1972). The structure is shown in Figure 4.1, and such an organization in the liquid state explains observed "memory" effects.

The structure proposal was derived from X-ray diffraction changes in connection with fat crystal melting; the occurrence of a diffuse peak in the long-spacing range with a position corresponding to that of triglyceride molecules in a bilayer with the chains in a liquid-like conformation. It was also based on the change of the X-ray scattering that was related to hydrocarbon chain length, and on relations with the L2-phases, discussed below.

Recently, however, a different structure was proposed (Cebula *et al.*, 1992). From neutron scattering studies on the trilaurin molecule, where different parts of the molecule were deuterated, they concluded that the molecules were oriented mainly in parallel, without forming a lamellar structure. They related their model to the general structure of the so-called nematic state of liquid-crystals. Their conclusions were later questioned, as the scattering data do not allow an unambiguous interpretation (Larsson, 1992a). A lamellar alternative cannot be ruled out by the experimental data.

Monte Carlo-simulation of the liquid state of triglycerides was recently started in order to get additional information on this problem (Engström, 1994). This computer simulation starts with the β-form conformation of the triglyceride molecule. After adjustment to the liquid state density, each cycle of the simulation procedure involves generation of new positions and calculation of the interactions between all atoms. The conformations are accommodated to the Bolzmann distribution, which means that conformations of low energy occur more frequently than those of high energy.

The preliminary results of this simulation study confirm the original proposal of a bilayer structure of the melt as indicated in Figure 4.1 Additional evidence is provided by the electron micrograph shown in Plate 4.2, discussed below.

B. LIQUID PHASES OF SIMPLE NON-POLAR LIPIDS COMPARED TO TRIGLYCERIDES

The reason why there is an order in the triglyceride melt is certainly related to the occurrence of three chains in the same molecule. Without an arrangement in layers, there is no way of obtaining a space-filling structure where the average chain direction of adjacent molecules is the same.

If we consider simple non-polar lipids, such as paraffins or long-chain alcohols, there are no such restrictions. In accordance with this in the X-ray scattering curve, there are no bands of comparable intensities to those observed in triglyceride melts. Furthermore the freeze-fractured electron micrograph of *n*-dodecanol shown in Plate 4.1 does not indicate any sign of order. Freeze-fracture is an informative technique for preparation of lipid samples for electron microscopy. When the sample is fractured in a frozen condition, it will open along planes with weak forces, and the methyl end groups are extremes in this respect.

A corresponding electron micrograph of a triglyceride melt (see Plate 4.2) shows a texture corresponding to the disordered bilayer structure discussed above, however.

C. THE L2-PHASE AND POLAR LIPIDS IN THE LIQUID STATE

When liquid-crystalline lipid-water phases are heated, they will ultimately undergo a phase transition ("melt") into a liquid state. In the phase diagrams shown in Chapter 3, this liquid phase was indicated by the L2-phase region. This phase has been analysed in monoglyceride systems (Larsson, 1979). It was concluded that the structure is similar to that of the lamellar liquid crystalline phase with a reduced size of the bilayer units, from "infinite" bilayers into curved bilayer units with a size of a few hundred Ångstroms in diameter. In other words the melting process means that the long-range order is lost, but the main features of the structure remain the same. Thus a medium range order arrangement of lamellar units is obtained. As the heating process gives a tendency for divergence of the hydrocarbon chains, it was also proposed that the polar regions were embedded in a continuous hydrocarbon chain matrix. This structure is illustrated in Figure 4.2.

The structure proposal was based on the following arguments:

a. Heating an L_α-phase through the transition into the L2-phase shows a diffraction pattern-versus-temperature diagram where the long-spacing (d(001)) of the $L_\alpha$phase is only slightly shifted, and the sharp line goes over to a diffuse band. The integrated intensity above and below the transition temperature is about the same, showing that the order in the L2-state also involves the whole bulk of the phase.

b. If it is assumed that the transition is a kind of melting of the L_α-phase,

Fig. 4.2. Proposed structure of the L2-phase in a monoglyceride-water system, after Larsson (1979). The black regions correspond to the polar head groups and the water layer in between. A section through such a structure is indicated by black regions; sometimes this section is parallel to the bilayer. The size of these discs is a few hundred Ångstroms.

so that the long-range order is lost although the structure over colloidal distances is the same, line-broadening analysis of the X-ray data can be performed. This gives a size for the scattering units of 300 - 500 Å, a value consistent with the rheological and optical properties of the L2-phase.

c. A plot of the position of the X-ray scattering band versus volume fraction lipid (c/1-c, where c is the lipid volume) gives a linear plot. This is always taken as a proof of a lamellar structure in the case of liquid-crystalline phases, and the same conclusion can therefore be drawn in the case of L2-phases. The bilayer thickness value obtained is also reasonable in relation to that in the L_α-phase.

d. Non-polar lipids do not exhibit this X-ray scattering characteristic.

Later studies of electron microscopy data were reported on the L2-phase in the system triglycerides/monoglycerides/water (Gulik-Krzywicki and Larsson, 1984). A freeze-fracture image of a monoglyceride-water L2-phase reproduced in Plate 4.3 shows differently oriented stacks of lamellae. The medium-range order of curved bilayers arranged in parallel is fully consistent with the model based on X-ray diffraction studies shown in Figure 4.2.

When the water content of the L2-phase is gradually reduced to zero, there is no discontinuity in the X-ray scattering pattern; the position corresponds to the bilayer thickness and the integrated intensity remains unchanged. The structure should therefore be the same. The freeze-fracture electron micrograph confirms the conclusion that the melt of polar lipids has the same structure as the L2-phase, as shown in Plate 4.4.

The liquid states of other polar lipids in the liquid state, such as galactolipids and phospholipids, show the same X-ray scattering characteristics as the monoglyceride melt discussed above. Furthermore they can always take up a limited amount of water into an L2-phase. The general conclusion is therefore that a polar lipid melt has a lamellar arrangement of bilayer units of finite size (a few hundred Ångstroms) in a short-range disordered/medium-range ordered structure.

D. MICROEMULSIONS

The term microemulsion is used for thermodynamically stable liquid phases consisting of oil, water, a surfactant and usually a "co-surfactant" (which is a small amphiphilic molecule, such as pentanol or octanoic acid). Microemulsions are usually prepared from a mineral oil and synthetic surfactants, and the criteria given for their identification refer to such systems (*cf.* Danielsson and Lindmann, 1981). In lipid systems, however, the oil is a triglyceride, which itself is surface active, and there is no need of a co-surfactant to form an L2-phase of a triglyceride oil, water and a polar lipid. It is obvious that this thermodynamically stable phase is a kind of microemulsion. The region of existence of such a microemulsion in a ternary system was shown in Figure 3.27.

The structure of the microemulsion in a system of sunflower oil monoglycerides/water/soybean oil was also discussed in connection with electron microscopy data of the monoglyceride melt and aqueous L2-phases mentioned above. A freeze-fracture electron micrograph of a monoglycerides/water/triglycerides mixture forming such a microemulsion is shown in Plate 4.5. If the X-ray diffraction is followed when the relative amount of triglyceride is increased, it can be seen that the lamellar order is steadily lost, and in parallel with this the maximum solubilization of water is reduced towards zero.

The microemulsion in lipid systems can thus be regarded as an aqueous liquid phase of lamellar units of water with polar head groups in a continuous hydrocarbon chain matrix. The units are a few hundred Ångstroms across. Such microemulsions consisting of water, a polar lipid and a non-polar lipid are of technical interest and will be discussed in Part 2.

E. THE L3-PHASE

The existence of an aqueous liquid phase with a high water content (about 70% (w/w)) in lipid systems was first demonstrated in an aqueous phase diagram of wheat lipids (Carlsson *et al.*, 1978). As this liquid could exist in equilibrium with water, it seemed natural then to term it L2. A liquid phase with the same characteristics was later observed in binary surfactant-water systems containing both L1- and L2-phases, and it was therefore given a separate name - L3 (*cf.* Andersson *et al.*, 1989). In binary systems, the phase can thus coexist with water on one side, and towards the lipid corner it usually coexists with the L_α-phase.

The structure has been suggested to be bicontinuous, consisting of "a highly connected bilayer extending in three dimensions" (Andersson *et al.*, 1989). The proposed structure is related to the cubic minimal surface structures we discussed in Chapter 3, although disordered due to weak interbilayer forces. By analogy with the discussion above relating the L2-phase to the L_α-phase, we can regard the L3-phase as similar to the cubic phase but the long-range order is lost.

The existence of an L3-phase coexisting with the H_I-phase on one side, as shown in Figure 3.21, is a new feature. An alternative explanation is the occurrence of a liquid-liquid miscibility gap, observed in aqueous systems of surfactants (Nilsson *et al.*, 1984). From the relationship between L3 and cubic phases, however, the existence of the L3-phase in this C_6C_{16}-PC-system can be related to the existence of a cubic phase in the same region of the phase diagram in the aqueous system of lyso-C_{10}-PC. Finally it should be mentioned that the L3-phase only has been observed in non-ionic surfactant-water system and in a few ionic surfactant-water systems containing salts (Andersson *et al.*, 1989).

Plate 4.1. Electron micrograph of a freeze-fractured electron micrograph of *n*-dodecanol, after Gulik-Krzywicki and Larsson (1984).

Plate 4.2. Electron micrograph of liquid triglycerides, kindly obtained from Wolfgang Buchheim, Kiel. A section through milk fat before crystallization is shown. The terraces have dimensions consistent with the bilayer thickness.

Plate 4.3. Freeze-fracture electron micrograph of a monoglyceride-water (92:8) L2-phase. Regions of bilayer stacks with different orientation and finite size, after Gulik-Krzywicki and Larsson (1984).

Plate 4.4. Freeze-fracture electron micrograph of a monoglyceride melt, after Gulik-Krzywicki and Larsson (1984).

Plate 4.5. Freeze-fracture electron micrograph of a sunflower oil monoglycerides/water/soybean oil (85:6:9) microemulsion, after Gulik-Krzywicki and Larsson (1984).

CHAPTER 5

MONOLAYERS IN SURFACE FILMS, BUBBLES AND FOAMS

As long ago as 1774, Benjamin Franklin described to the Royal Society how a drop of a vegetable oil spread over the surface of a water pond, as was evident from damping of small surface waves. The remarkable spreading behaviour into what was assumed to be a monomolecular film on the water surface gave an experimentally based understanding of the size of these molecules. A detailed molecular picture of lipid surface films was then presented at the beginning of this century by Langmuir (1917). The presence of surface films on the surface of the sea can sometimes be observed by the damping of the waves ("sea slicks"), and such monolayers usually consist of both lipids and proteins. In this chapter, we will only consider lipid surface films; lipid-protein interaction is considered in the next chapter.

A. FORMATION OF MONOMOLECULAR SURFACE FILMS AT THE AIR/WATER INTERFACE

In his experiments, Langmuir spread simple lipids on the surface of water between two barriers in a trough. One barrier was used to compress the film, and the other was used as a sensor of the film pressure. In this way, the film pressure (Π) versus molecular area (A), Π-A isotherms, could be recorded. Monolayers of lipids occur in different two-dimensional states, the gaseous, the liquid and the solid states. The terminology used is G (gaseous), LE (liquid-expanded), LC (liquid-condensed) and S (solid). It does not, however, take into account polymorphism of the solid phases, and is therefore confusing (we can call the solid phases S_1, S_2 *etc.*). It is quite clear now that the LC-phase has a crystalline structure. Its higher compressibility than "S-phases" is due to the tilt of the hydrocarbon chains. It is also clear that the chain conformation in the LE-phase is the same as in the lamellar liquid-crystalline phase, as shown by compressibility and molecular cross-sectional area.

When we spread a monolayer with molecules far apart, they will exhibit the gaseous state. On compression we can first form a condensed liquid

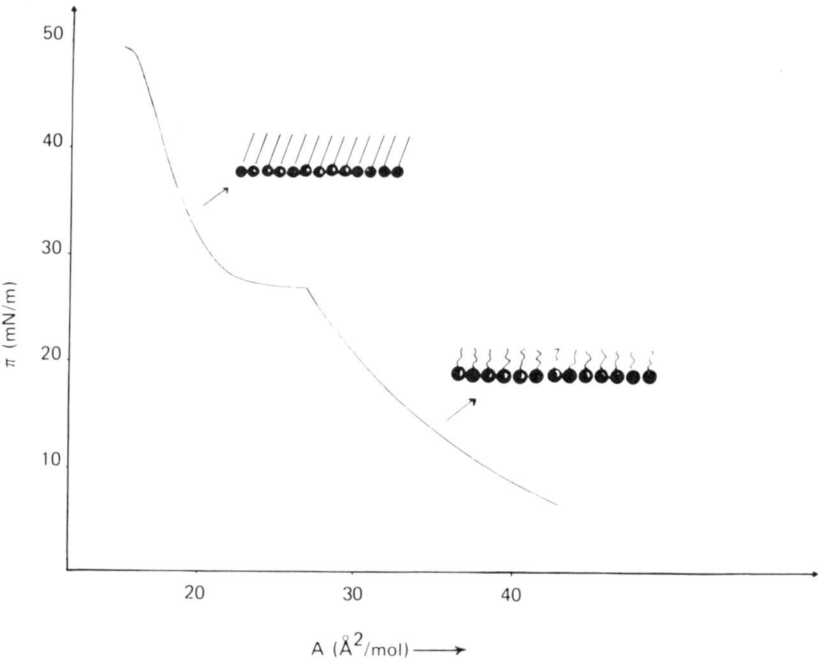

Fig. 5.1. Pressure-area (Π-A) isotherm of 1-monomyristin at 25°C. The structure of the two monolayer phases I and II are also indicated, after Krog *et al.* (1985).

phase and then solid phases. Sometimes there is a direct condensation from a gaseous to a solid state. Two-dimensional pressure-temperature phase diagrams can be constructed by recording Π-A isotherms at different temperatures.

The appearance of different phases in monolayers can be observed by adding a fluorescent probe, which will partition differently between the phases, a technique introduced by McConnell (*cf.* von Tscharner and McConnell, 1981). They have demonstrated beautifully how two-dimensional crystals of phospholipids are formed on a water surface, exhibiting shapes reflecting the antipode character of the molecules.

Another interesting macroscopic feature demonstrated in two-phase regions of gaseous-liquid phases is the occurrence of a two-dimensional "foam" (Moore *et al.*, 1987). When the film is expanded, the gas "bubbles" grow from a circular shape and start to close-pack to give polygonal structures with a thin layer of liquid in-between. Such two-dimensional "foams" exhibit an interesting geometry, which also provides information on foam structure in general (*cf.* Stavans and Glazier, 1989).

1. Monolayers of Polar Lipids

The monolayer behaviour of 1-monomyristin is shown in Figure 5.1, and the corresponding phase diagram, obtained from Π-A isotherms

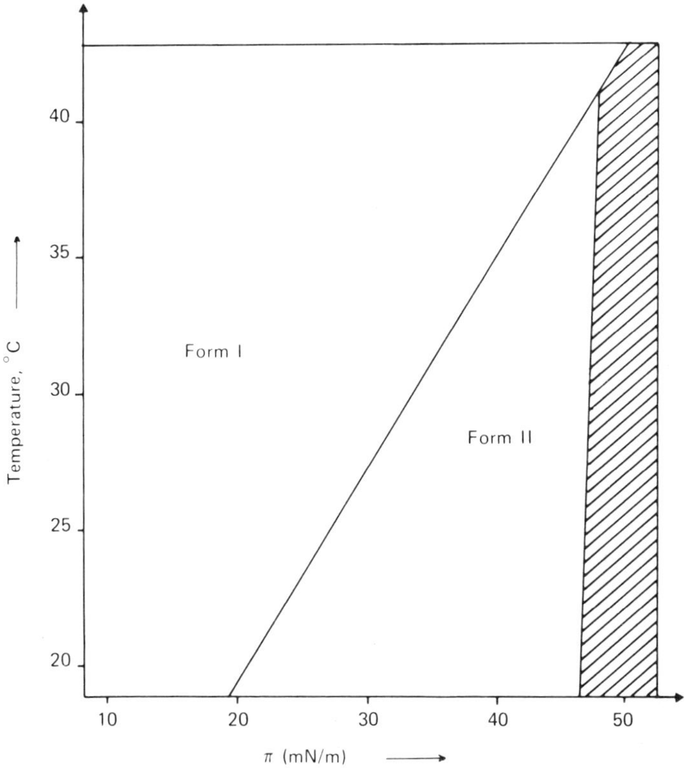

Fig. 5.2. Phase diagram of 1-monomyristin monolayers. The shaded area corresponds to collapse of the monolayer, after Krog *et al.* (1985).

recorded at different temperatures, is given in Figure 5.2 (Krog *et al.*, 1985). The first phase, which is formed during compression from the gaseous state, form I, has the same structure as the L_α-phase in three dimensions. At about 25 mN/m, there is a transition into another monolayer phase, form II. The equilibrium between the two phases, evident as a plateau with constant pressure, shows that it is a first order transition. The cross-sectional area per molecule at the transition is 28.5 Å^2 in form I and 22.5 Å^2 in form II. The cross-sectional area of form I is consistent with the liquid-chain conformation of L_α-phases, and the area of form II is the same as that of monoglyceride crystal forms (*cf.* Chapter 2).

The most interesting feature of this phase diagram is the range of existence of the two phases. Above 42°C there is no phase with crystalline chains. This temperature is in agreement with the transition from the L_α-phase to the gel-phase in three dimensions. Thus there is a close relation between the phase behaviour in three-dimensional aqueous systems and in monolayers. 1-Monoelaidin, which was also examined, shows the same good agreement between monolayer and bulk phase transitions. The following simple relations have thus been demonstrated:

a. If a polar lipid forms an L_α-phase with water in three dimensions, it will give only one monolayer phase with chains in a liquid state at the same temperature.
b. If a polar lipid at the actual temperature forms a gel-phase with water, the monolayer will first exhibit a condensed monolayer with the chains in the liquid conformation, then at further compression a first order transition into a monolayer in the solid state.

There are numerous reports on the monolayer behaviour of membrane lipids. They usually form an L_α-phase with water, and therefore the simple monolayer behaviour described above. On the other hand, synthetic phospholipids, for example dipalmitoyl-PC, exhibit different monolayer phases in the solid state (Albrecht *et al.*, 1981). Basically, they follow the second relationship given above. Among other membrane lipids, which have been studied, it can be mentioned that cerebrosides behave like phospholipids (Abrahamsson *et al.*, 1972). A special feature is the existence of solid phases, which do not collapse until a film pressure of about 60 mN/m.

2. *Macroscopic Structure in Lipid Monolayers*

Extensive studies of phospholipid monolayers using fluorescence microscopy have given details of domain shapes in the solid-liquid coexistence region (*cf.* Weis, 1991). Thus equilibrium domain shapes have been observed at very slow compression/expansion when the Π-A isotherm becomes reversible. The two factors determining the shape are the solid/liquid line tension and the electrostatic forces at the interface. When the line tension dominates and the solid domain is isotropic in the two dimensions, a circular shape of the domain would be expected. The surface potential of a monolayer phase is constant, and at the boundary between a liquid monolayer and a solid domain, a discontinuity in surface potential will occur. Any asymmetry in the molecular orientation in the plane of the domain will be reflected by the shape of the domain, as the line tension as well as the electrostatic forces will vary in different directions. Furthermore the repulsive forces between different domains will result in an ordered superstructure, and dimyristoyl phosphatidic acid has been observed to form a hexagonal array of domains, for example (Lösche *et al.*, 1988).

3. *Triglyceride Monolayer Behaviour*

The most important interfacial property of oils consisting of triglycerides is the surface activity of the molecules. Thus, they will spread

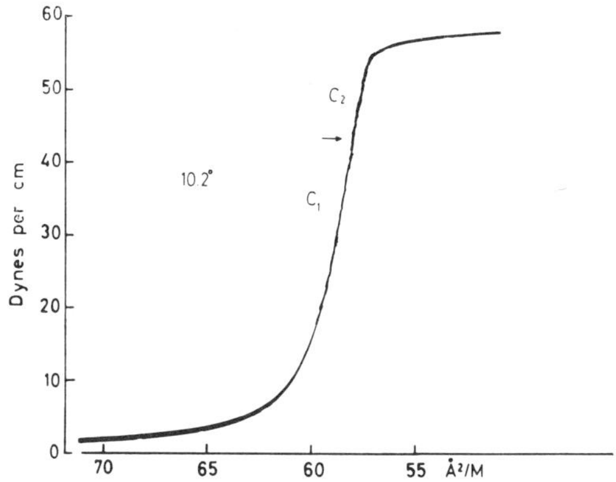

Fig. 5.3. Tripalmitin Π-A isotherm at 10°C, after Burch *et al.* (1968). The two solid phases are called C_1 and C_2.

on water in order to reduce the surface free energy. Compared to other non-polar lipids forming monolayers as described in the next paragraph, the triglyceride molecule in contact with water will have its three chains directed up from the water, in order to allow interaction between the polar part of the molecule and water. This means that there is a difference in molecular conformation of the molecule compared to the bulk structure of triglycerides.

Monolayers of monoacid triglycerides were studied first by Dervichian (1939), who demonstrated the main monolayer properties. A more detailed study was later reported on tripalmitin (Burch *et al.*, 1968). Above 44°C, no solid phase was seen, and this means that the chains are melted at all pressures. In the condensed monolayer region, there were shifts in the slope of the Π-A isotherms indicating three different phases, and it was proposed that they may differ in hydrocarbon chain close-packing. A characteristic Π-A isotherm is shown in Figure 5.3. The interesting features of this isotherm are the steep rise in pressure, when the gaseous film is transformed directly to a solid-condensed phase, and the phase transition in the solid state.

Studies of mixed chain triglycerides have been reported by Fahey and Small, involving 1,2-dipalmitoyl-3-acyl-glycerols with 2 to 8 carbon atoms in the 3-position (Fahey and Small, 1986) and with 10 to 16 carbon atoms in this position (Fahey and Small, 1988). The short hydrocarbon chain members (up to C_6) exhibit a most remarkable conformation on the water surface of immersion of the short chain into the water. The shortest member, C_2, behaves just like dipalmitin with the two long chains only above the water in the liquid as well as the solid monolayer. In the case of longer chains (C_3 to C_6), the short chain is immersed in the water only at pressures high enough to give solid monolayers. The equilibrium spreading pressures (ESP) were determined just above the chain melting and found to go from 20.6 to 14.8 mN/m in the series C_2 to C_8. The

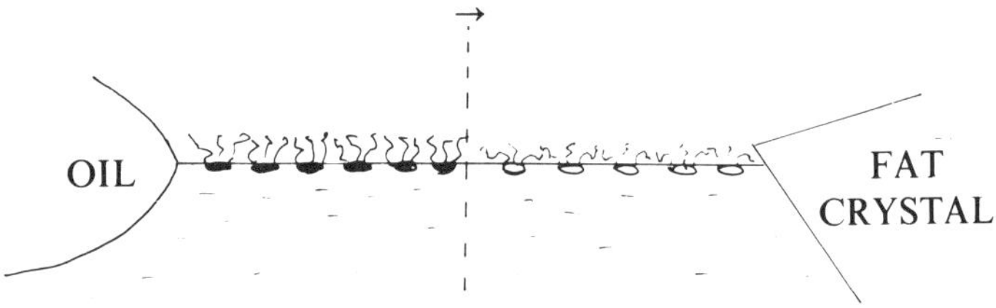

Fig. 5.4. Illustration of the competition between triglyceride molecules spreading from a liquid and from a crystal. The movement of the border between the two types of films is indicated (provided that a monolayer around the crystal and droplet is first obtained). The triglyceride molecules from the oil are indicated by filled polar heads and those from the crystal by open heads.

ESP-value corresponds to the equilibrium between crystals (or oil) and the monolayer, and it can be obtained by just depositing the bulk phase on the water surface.

The longer monolayers show ESP-values successively reduced to 10.5 mN/m when tripalmitin is reached. At chain lengths above C_{10}, there is first a transition directly to a solid phase from the gas phase (as in Figure 5.3), and at higher temperature a liquid monolayer phase exists before the solid phase is obtained.

4. Competitive Spreading of Monolayers from Fats and Oils

The lesson to learn from the monolayer studies described above is that the presence of triglycerides as an oil or as crystals in any complex mixture containing non-polar/polar interfaces will result in formation of a triglyceride monolayer at the interface, providing that no other amphiphilic molecules with a higher spreading pressure is present. In the case of a plastic fat, the higher ESP value of the liquid oil will remove all triglyceride crystal molecules from the interface, as indicated in Figure 5.4.

A practical example of spreading of a triglyceride film is the addition of some margarine to boiled pasta. A monolayer film of triglyceride oil is evenly distributed on the surface because of the spontaneous spreading described above.

The kinetics of the spreading process can vary appreciably as will be described below. A monolayer is spread within a second when a liquid triglyceride is dropped on the surface balance trough. In contrast, a crystal of a triglyceride forms monolayers very slowly, and only those with very short chains show a film spreading behaviour of practical significance.

In order to know the kinetic of spreading when fats and oils compete at the interface, it is necessary to study different triglycerides in different crystal forms. Some new results on this problem are reported below.

The liquid states of trioctanoin, tricaprylin and triolein were compared at 35°C. The ESP-values recorded were 18.6, 18.0 and 12.0 mN/m respectively. Trioctanoin at 23°C gave a value of 20.6 mN/m. At air/water

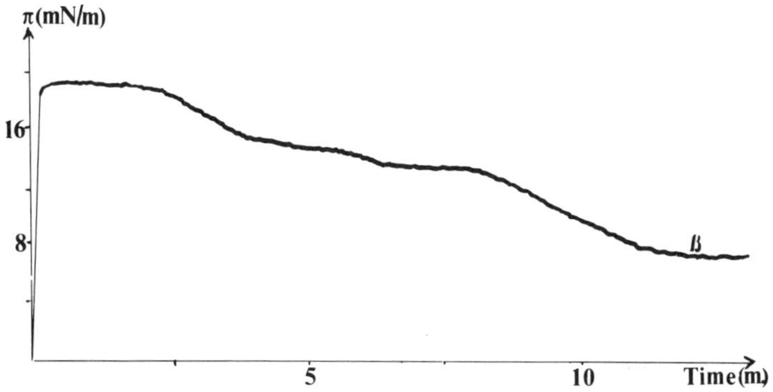

Fig. 5.5. A direct recording of monolayer pressure versus time of a droplet of liquid trimyristin deposited on a water surface at 35°C.

interfaces stabilized by protein monolayers with a common film pressure of about 20 mN/m, it is obvious that short-chain triglycerides in the liquid state may penetrate the protein monolayer. Foam destabilization due to the presence of oils is a well-known phenomenon. Acetylated C_{16}/C_{18}-monoglycerides, a triglyceride type used as a functional food additive, showed an ESP-value of 29.4 mN/m at 23°C in the liquid state. Such a film will squeeze out a protein monolayer from the air/water interface..

The presence of triglycerides in the solid state with regard to monolayer formation is related to the polymorphic behaviour, the equilibrium between the monolayer state of the triglyceride molecule and the actual crystal form. Crystals of the β-form of tridecanoin on a water surface at 23°C gave an ESP-value of 14 mN/m within minutes, whereas trilaurin showed a value of about 6 Å after one hour, which seemed to be the β-form ESP. The spreading was very slow, however, with almost no pressure effect in the first half hour. When trimyristin β-crystals were deposited on the water surface, no film pressure could be observed within the experimental time (two hours).

When β'-crystals of triundecanoin were deposited on the water surface, a remarkable behaviour was observed. The pressure increased to a plateau value of about 10 mN/m within a few minutes, and after about ten minutes it increased to 15 mN/m. The most probable explanation of this behaviour is a polymorphic transition in the monolayer, as is seen also in Figure 5.5.

In an effort to obtain an ESP-value of the α-form, trimyristin in the liquid state (just above its melting point) was deposited on a water surface at 35°C. As this temperature is below the α-melting point, the liquid triglyceride must be expected to transform directly into the α-form at the water interface (see Figure 5.5). Thus the high ESP-value first obtained of 19 mN/m should be that of the α-form. When the pressure then goes down, an intermediate level is reached before the final plateau value is obtained. From the phase transition of the bulk phase of trimyristin, we should expect that a β'-form will occur before the β-form finally is obtained. The curve might therefore reflect ESP-values of the α-, β'- and β-forms.

5. Monolayers of Other Non-Polar Lipids

Free fatty acids often give a complicated behaviour when they occur in lipid systems, as they form aqueous phases when they are ionized, whereas they behave like non-polar lipids at neutral and lower pH. In the undissociated state, they can form various solid phases on a water surface. Pressure-area isotherms, which are characteristic for long-chain saturated fatty acids, are illustrated in Figure 5.6.

The pK-value in a monolayer is quite different from that in the bulk phase, and the packing environment in condensed monolayer phases results in a pK_a-value in the range 6-8. There is a direct transition from molecules in the gaseous state to a solid phase ("L_2") with a structure with about the same angle of tilt as a crystalline phase (B or C). At higher pressures, solid monolayers with vertical chains (phases LS and S) can be formed. If we stop the compression in this region and wait, the pressure will go down to the equilibrium spreading pressure, and this value is the same as that of the phase "L_2" at its highest pressure. The same pressure is also seen after collapse. This shows that fatty acid crystals on water will form stable monolayers when the structure resembles that of the crystals.

The monolayer behaviour of 1,2-dipalmitin was reported by Fahey and Small (1986). At room temperature, there is a direct transition from the gaseous state to a solid monolayer. The equilibrium spreading pressure is 31.7 mN/m, just above the chain melting point.

Simple esters on the water surface have been studied in detail by Lundqvist (1978), who revealed two remarkable features:

- the monolayer polymorphism appeared to be identical with the polymorphism in the solid state; even the alternation in thermal transitions coincided.
- butylstearate or amylstearate form close-packed monolayer phases, which means that the polar group is well above the water surface.

In a study of a diglyceride monolayer, a similar behaviour was seen. When a monolayer with a molecular cross-sectional area corresponding to two chains is compressed further, a conformation is obtained with a molecular cross-sectional area corresponding to one chain (Larsson, 1973a). This means that the glycerol group is located in the centre of the bilayer, which is one of the two conformations occurring in the solid state (Chapter 2).

6. Mixed Monolayers

As we learned about mixtures of lipids in three dimensions, lipid phase behaviour in monolayers shows molecular solubility or phase separation

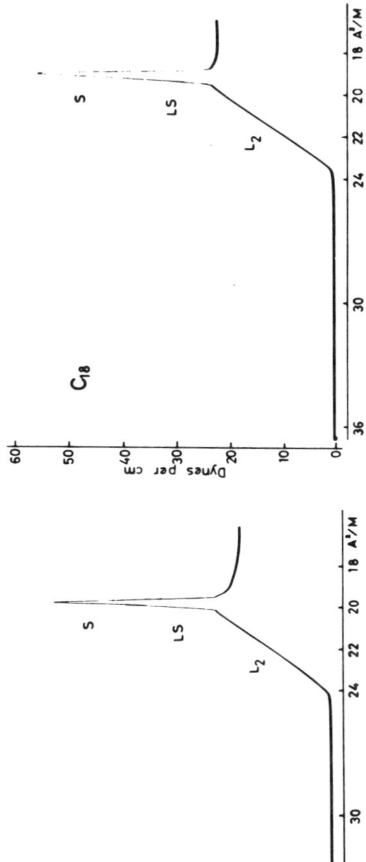

Fig. 5.6. Pressure-area (Π-A) isotherms of palmitic and stearic acids (on 0.01N HCl-solution) at room temperature, after Lundqvist (1978).

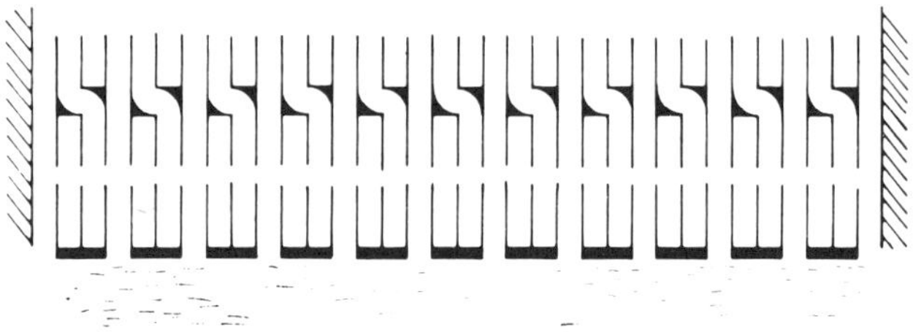

Fig. 5.7. The formation of a triple layer of tripalmitin on the water surface (Larsson, 1973).

depending on the interaction between the components. By recording Π-A isotherms of mixtures between two lipid components, it can be seen whether the curve is an average of those of its components and is related to the amount of each component. In this instance, there is molecular mixing. Alternatively the characteristic features (phase transition and collapse point) of the isotherm of each component are seen, demonstrating that islands of one component are dispersed in a continuum of the other. However, this picture is complicated in practical examples by differences in mutual solubility in different states. In gaseous films, there is always complete solubility. In the liquid state, there may be phase separation, and this is much more common in the solid state.

An interesting study of a biologically important system concerns mixed films of dipalmitoyl-PC and cholesterol (Subramaniam and McConnell, 1987). The appearance of the phases was followed by a fluorescent lipid probe, and coexisting fluid-fluid domains were visualized. At a cholesterol concentration of 30 mol%, fluctuations in domain shapes were observed, indicating a critical point of immiscibility.

7. Multilayers on Water and on Solids

When monomolecular films of lipids on water are compressed beyond the monolayer collapse point, multilayers can sometimes be formed as is evident from the Π-A isotherms (Larsson, 1973). Thus fatty acids and methyl esters can show Π-A isotherms corresponding to 3, 5 and sometimes 7 monolayers. This corresponds to the existence of lipid bilayers above the monolayer in contact with water. It seems obvious that a fatty acid bilayer cannot exist in a water surface. Only when there is an odd number of monolayers is a hydrophobic surface exposed towards the air. Thus folding of bilayer units above the monolayer takes place. Ethyl esters, however, can form multifilms with 2, 3, 4, 5 and 6 monolayers. This reflects the solid state behaviour of ethyl esters, with no tendency for bilayer formation, as described in Chapter 2. The formation of multilayers seems to be related to the occurrence of a solid monolayer with the same molecular disorder as the α-form in solids (Larsson, 1973b). An example of multilayer formation is shown in Figure 5.7. The conformation of the bilayer above the monolayer in contact with water has the same type of structure as triglycerides in the solid state (*cf.* Chapter 2).

So-called Langmuir-Blodgett (LB) multilayers can be deposited in a solid support (*cf.* Blodgett (1935)). For example, by moving a solid with a surface which is hydrophobic down through a monolayer of fatty acids into the water phase, it will be covered by a monolayer (provided that the conditions for monolayer transfer are fulfilled). The driving force is the reduction of the surface energy. Such transfer processes are simplest when the lipid molecules are in a solid phase, preferably an α-form. When the solid is moved up from the water phase through the monolayer, another monolayer will be transferred, making the surface hydrophobic in air. By repeating this process, LB-layers with desired numbers of monolayers can be produced. There are interesting technical applications of LB-films in electronics and in optics, that have been based on the ingenious work by Kuhn and Möbius (1971).

Polar lipids that can stabilize foams effectively should be water-soluble. Therefore it is only the more-strongly polar lipids, like lysophospholipids and soaps, that form foams. The reason is the need for water solubility in order to supply monomers to the expanding gas/water interface quickly enough.

In the following text, we will analyse the forces involved in stabilizing bubbles and foams.

B. BUBBLES AND FOAMS

Because of the amphiphilic character of lipid molecules, they can stabilize gas bubbles generated in a water phase. The gas/air interface is thus stabilized by a lipid monolayer of the same structure as was discussed above. When gas bubbles have a long lifetime, it is possible to increase their volume in the water phase. When spherical bubbles reach a volume fraction of about 0.75, the gas cells become polyhedral in shape, and a foam is obtained ultimately. The foam lamellae tend to meet at angles of 120°, and a cross-section of a lipid foam is shown in Figure 5.8.

A "free" gas cell covered by the same surface monolayer on each side will of course become spherical, *i.e.* a soap bubble.

1. Foam Structure

If we examine any foam, it can be seen that the size of the gas cells varies, and so does the shape of the polyhedra. From statistics on the average number of faces of each polyhedron, it has been found that most polyhedra have 14 faces (Williams, 1968). This is a most interesting polyhedron, which was described some time ago by Kelvin as that with the smallest surface-to-volume ratio. It consists of 6 square and 8 hexagonal faces.

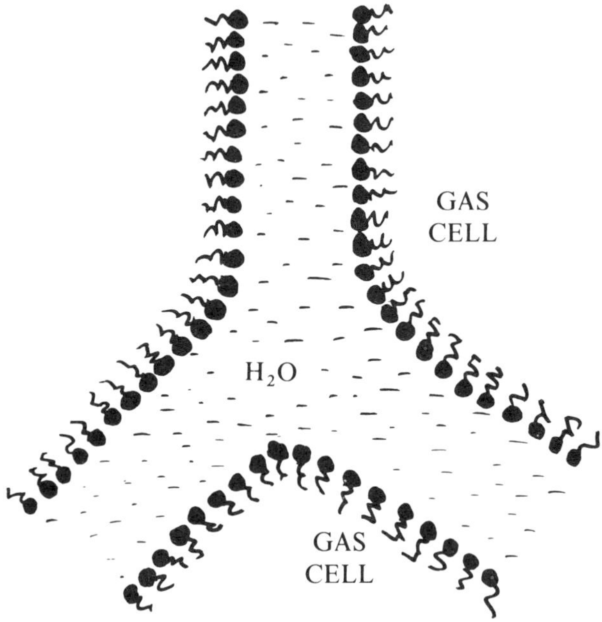

Fig. 5.8. Cross-section through a foam, showing the lipid monolayer at the gas/water interface.

2 General Factors behind Foam Stability

The stability of a foam is related to the following factors:

- mechanical properties of the lipid monolayer
- repulsive forces between the monolayer on each side of the lamellae
- viscosity of the water phase, influencing drainage.

Drainage is caused by gravity, and also by the pressure in the so-called Plateau borders (where the lamellae meet). The pressure in the Plateau border is smaller than that in the central part of the lamellae, as the lamellae are almost planar, whereas the Plateau border is curved. Thus the liquid within each lamella will flow towards the Plateau border, and then there will be a flow in the connected systems of Plateau borders. Against this flow, repulsive forces act between the surface monolayers, *i.e.* usually electrostatic repulsion.

Let us consider a single foam lamella. Equilibrium occurs when the hydrostatic pressure is balanced by a so-called disjoining pressure. The gravitational field defines the hydrostatic pressure, and the disjoining pressure is due to repulsive forces between the opposing film surfaces. In addition to this, there is an obvious effect of viscosity on the drainage of the foam.

Surface film viscosity, elasticity and the Marangoni effect are the most important factors in determining the stability of a foam. The Marangoni effect is due to differences in surface tension between different points at the surface of the foam lamellae. Assume that a foam lamella starts to burst. When a hole is formed by local thinning of the lamella, the surface area will increase. Because of the low diffusion rate of the surface active molecules, the surface tension at the hole will be higher than that of the surrounding part of the lamella. This spot with a larger surface tension will therefore obtain monolayer molecules from the surroundings, which due to viscosity will also drag liquid from the inner part. The hole will therefore be closed.

3. Effects of Electrolytes on Bubble Stability - The Hydrophobic Force

A remarkable effect of common electrolytes on gas bubble stability in water has recently been described (Craig *et al*., 1994). Bubble coalescence was thus analysed in various electrolytes at different temperatures. Air bubbles introduced into pure water have lifetimes below a second, whereas the salts give lifetimes of up to 10 seconds (which can be compared with surfactants with bubble lifetimes up to hours). When the bubble coalescence versus concentration was followed, there was a steep increase in stability above a critical concentration. In the case of sodium chloride, this concentration is about 0.1 M, whereas divalent salts show a much lower critical concentration (0.04 M for calcium chloride).

The authors present a most interesting theoretical discussion of this phenomenon, which "no known forces of colloid science can explain". They propose that the phenomenon is related to the hydrophobic interaction (Israelachvili and Pashley, 1982). This long range attractive force was observed at distances between hydrophobic surfaces in water of several hundred Ångstroms. It is 10-100 times larger than the corresponding van der Waals forces, according to experimental measurements.

It seems likely that the observation on bubble stabilization by electrolytes can explain the hydrophobic effect. There is evidence that dissolved gases close to saturation in water may induce microbubble formation at a hydrophobic surface (Craig *et al*., 1994). The measured hydrophobic force is reduced in strength and range by the addition of salts, and the effect is proposed to reflect the salt effects on bubble stability. The mechanism underlying bubble stabilization is concluded to be due to water structuring rather than any electrostatic effect, when comparison is made with bubble stabilization by different sugar molecules. The results are suggested to be relevant even to the evolution of life; the salt level in our body fluids is the minimum for maximum bubble stability.

4. Micellar Lifetime, Bubble Dynamics and Foamability

When a foam is produced, by for example gas injection into a micellar solution, the surfactant (lipid) molecules adsorb at the gas/water interface.

The diffusion time of the surfactant molecules is much shorter (about 10^{-4} s) than the micellar lifetime (about 0.6 s), according to estimations in sodium dodecylsulphate (SDS) solutions (Oh and Shah, 1991). The lifetime of a micelle is therefore the rate-limiting step, as new monomers are only formed by disintegration of micelles.

In a recent work, a direct relationship between foamability and the micellar lifetime was demonstrated (Oh and Shah, 1991). In solutions of SDS and SDS-hexanol mixtures, the micellar relaxation time was determined. At a concentration of 200 mM SDS in the pure SDS-solution, the maximum in micellar lifetime was observed. This concentration is probably related to the so-called *second critical micellar concentration* (CMC). This change in the micellar solution is characterized by formation of cylindrical micelles, which start to grow at the expense of the spherical ones. It was found that the foam had its minimum stability at the concentration with the maximum in micellar lifetime.

When foam is produced by injection of gas bubbles, the bubble dynamics will determine their size and therefore the structure of the foam. If we consider a capillary where the bubbles enter the solution, the bubble will detach itself from the tip when the buoyancy force balances or exceeds the surface tension of the contact circle between the bubble and the tip. Thus, the bubble size decreases with the surface tension. Above the CMC of surfactant molecules, the surface tension is constant, however. On the other hand, the sizes of bubbles produced have been found to vary with surfactant concentration. This dynamic effect is again related to the micellar stability (Oh *et al.*, 1992). Thus a maximum in bubble radius was observed when the micelles were most stable during bubble generation at a constant gas flow.

Salts can influence foam stability in several ways. The addition of a salt will lower the CMC of the surfactant, and a lower monomer concentration in the solution will have a negative effect on bubble dynamics and foamability in accordance with the discussion above. The electrostatic repulsion between the monolayers of surfactants on each side of foam lamellae will be reduced by salts in the case of ionic surfactants. This should be expected to destabilize the foam, provided that the monolayer structure of the surfactant is not changed.

5. *Destabilization of Foams*

There are many applications of foams, and sometimes foaming must be reduced. Inhibition of foaming is therefore also a technology we may need. Destabilization mechanisms of the foam structure will reduce foaming. The effect of salts on foam formed by ionic surfactants results in destabilization usually, as explained above. A common antifoaming agent used in technical applications is 2-ethylhexanol. The effect is due to a strong reduction in the surface viscosity of the surfactant monolayer in the foam surface. Another mechanism is formation of separate monolayer regions with poor mechanical properties, so that the foam lamellae are ruptured. An example is protein foams, which can be destabilized by fatty acids, such as oleic acid. The protein-stabilized foam of beer can be destabilized in this way when the brewing process has resulted in "over-foaming" of the beer. The fatty acid molecules will form monolayer zones which do not mix with the protein monolayer. Even if these zones only cover a minor part of the surface of foam lamellae, it is enough to open it.

An interesting new mechanism of foam destabilization has been reported recently (Blute *et al.*, 1994). They examined a series of organic electrolytes, which form counter ions to SDS-stabilized foam. Tetramethylammonium bromide (TMAB) and its homologues, tetraethyl- (TEAB), tetrapropyl- (TPAB), tetrabutyl- (TBAB) and tetrapentyl- (TPAB), were compared with sodium bromide with regard to the half-life of the foam produced. The half-time values were drastically reduced by the addition of these organic bromides, and showed the order TEAB $>$ TBAB $>$ TPAB (the same molar concentration of sodium bromide had a minor effect). It was shown that the destabilizing effect is due mainly to the bulky geometry of the organic ion, which gave an increased area per surfactant molecule at the air/water interface. The reduction in molecular area results in a decreased surface viscosity of the monolayer.

The destabilizing effect of these bulky organic ions was observed also in foams stabilized by non-ionic surfactants. Obviously the organic cations penetrate the monolayer at the surface of foam lamellae and destabilize them by reducing their viscosity.

6. Foams Containing Proteins and Emulsified Oils/Fats

As protein molecules are amphiphilic, they will tend to adsorb and orient at gas/water interfaces in order to reduce the surface free energy. For this reason they can also stabilize foams, and egg white is a well-known example. Another type of foam with which we all are familiar is whipped cream. In this case, the fat globules stabilize the foam by aggregation within the foam lamellae. These types of foams will be discussed in detail in the chapter on foods. Although such foams are much more complex, because of the presence of many components, the basic principles of stabilization and destabilization mechanisms described above can still be applied.

CHAPTER 6

LIPID INTERACTION WITH PROTEINS AND OTHER POLYMERS

The main explanation for the interaction between lipids and proteins is their amphiphilic character. They therefore accumulate and orient at interfaces, or associate in an aqueous environment without interfaces. The common feature of possessing polar and non-polar regions in the same molecule can sometimes lead to such strong interactions that these are regarded as complex formation. We will use the term interaction in this context to describe non-covalent association between lipids and proteins on the *molecular level*. There is sometimes an interaction between these components only on the *colloidal level*, however. One example is emulsification of a triglyceride oil by polar lipids or protéins, when all of the triglyceride molecules in the core of the oil droplet do not interact with the emulsifier. We will discuss this situation in Chapter 8. Only molecular interactions will be considered in this chapter.
chapter.

A. SOME NEW ASPECTS OF INTERFACIAL PROPERTIES OF PROTEINS

The reader is assumed to have a basic knowledge of protein structure and the description of the organization of the peptide chain as defined by the primary, secondary and tertiary structure levels. There are often separate globular regions within a molecule that behave as units and are called *domains*. In aqueous phases, the protein molecules are sometimes associated, and the assemblies of associated units are classified as the *quaternary structure*. The forces determining the aqueous structure, from folding of the peptide chain to the quaternary structure are the same as those responsible for the assembly of lipid molecules in the variety of phases discussed in Chapters 3 and 4. The occurrence of covalent links within or between peptide chains is the only major difference. A simple model of a globular protein in solution is therefore rather like the micellar structure; the polar amino acid groups are directed towards water, whereas non-polar groups tend to avoid water contact by forming the core. *Denaturation* is an unfolding process, often induced by increased thermal mobility, and the resulting exposure of hydrophobic segments of the peptide chain usually results in *aggregation.* If the aggregates form a three-dimensional network through the water matrix, the structure is described as a *gel.*

A new state, called the *molten globule state*, of proteins has recently been described (Hirose, 1993). The term was proposed to describe the non-native state of cytochrome c, with a compact conformation of the polypeptide chain and an enthalpy at transition from the native form rather like the melting enthalpy of a solid. Serum albumin and conalbumin are transformed into this state when the disulphide bridge is reduced. Many other proteins, such as α-lactalbumin, β-lactamase and apomyoglobin, adopt this state at acidic pH. At very low pH, total denaturation occurs. The presence of anions can then induce the molten globule state conformation.

The molten globule state has the following features

- the secondary structure is more or less intact, whereas the tertiary structure is lost
- the hydrodynamic volume is slightly expanded compared to the native state
- the binding of hydrophobic groups is increased, indicating exposure of hydrophobic regions

The molten globule state interacts more efficiently with lipid-water structures than the native state of the proteins. The molten globule state has therefore interesting functional properties in stabilizing foams and emulsions (Hirose, 1993). It also has important biological functions, as has been shown in connection with insertion of proteins into membranes (van der Groot *et al.*, 1991).

B. LIPID-PROTEIN INTERACTION IN COLLOIDAL PARTICLES

The general features of the interaction between water-soluble lipids (or surfactants) and proteins have been reviewed by Tanford (1980). Most model studies concern the interaction between globular proteins and sodium dodecyl sulphate (SDS). The non-specific cooperative binding of SDS usually leads to unfolding of the peptide chain, resulting in random coil and α-helix conformations. Hydrophobic side-chains will be covered by SDS molecules, making the exterior hydrophilic. At a high amount of SDS, micellar clusters seem to occur along the flexible random coil chain (Maltice *et al.*, 1976).

Non-ionic surfactants usually have little or no interaction with globular proteins (apart from specific interactions, as in serum albumin discussed below). They can interact with membrane proteins, however, making them water-soluble.

Cationic surfactants can show a cooperative interaction with globular proteins, but with less affinity than shown by anionic surfactants (Tanford, 1980).

1. Interaction of Serum Albumin with Fatty Acids

Free fatty acids are transported in plasma by serum albumin, which contains at least five fatty acid binding sites, three of which are high-affinity sites. The binding mechanisms in the individual binding sites have been determined by an elegant ^{13}C-NMR study by Hamilton (1992). The average lifetime for a long-chain fatty acid molecule in a high-affinity site is $>$66 ms, whereas short-chain members such as octanoic acid have a lifetime of only a few ms. This is mainly due to exchange of fatty acid molecules between different binding sites, whereas the dissociation of a fatty acid molecule from albumin has a time-scale of seconds. The dissociation is the rate-limiting step in the delivery of fatty acids to membranes of various tissues. In this molecular complex, hydrophobic pockets for the ligand are involved. In addition, water-insoluble molecules other than fatty acids, for example the pigment bilirubin or drug molecules, are carried by albumin as molecular complexes.

2. Lipoproteins

Large colloidal aggregates, the lipoproteins, transport other lipids in the blood circulation. Their sizes are directly related to the relative amounts of lipid, and they are classified in order of increasing size as high-density lipoproteins (HDL), low-density lipoproteins (LDL) and very-low density ones (VLDL).

HDL have the important function of removing cholesterol from cell membranes via the circulation. Apolipoprotein A-I encapsulates a disc-shaped region of membrane containing a few cholesterol and phospholipid molecules. The enzyme lecithin-cholesterol-acyltransferase then transforms the cholesterol to cholesterol esters, which form the core of the HDL particle when it emerges from the membrane. The HDL-particle is perhaps best described as a molecular complex.

LDL, VLDL and chylomicrons are the main carriers of lipids (in the form of cholesterol esters and triglycerides) to various tissues. These particles resembles emulsion droplets, with a diameter of a few hundred Å in the case of LDL and up to several thousands Å in the case of VLDL and chylomicrons. The apoprotein E-3 occurs on the surface of VLDL and chylomicrons.

An extensive conformational analysis of the apoproteins A-I and E-3 was recently reported (Nolte and Atkinsson, 1992). Both proteins are rich in helical regions, and a very important feature responsible for lipid interaction is that one side of the helices is hydrophobic and the other is hydrophilic. The balance between hydrophobic/hydrophilic regions determines the degree of penetration into the lipid core, and possibly even enzyme activation.

The LDL-particle has been studied extensively, because of its significance in cardiovascular diseases, a topic which will be discussed in Chapter 14. The interior of the LDL-particle contains about 1000 cholesterol ester and triglyceride molecules, with cholesterol esters predominating. Based on a general knowledge of lipids at interfaces, it is believed that the surface consists of a monolayer of phospholipid and free cholesterol molecules with apoprotein B embedded. The fatty acid composition is remarkably constant (88:12 of $C_{18:2}$:$C_{20:4}$), and this may indicate a specific internal organization. In view of the existence of cholesterol ester liquid-crystalline phases (*cf.* Chapter 3), it seems possible that the core has the same structure as the blue phase with one subunit forming the LDL-core.

The uptake of LDL through membranes in various tissues was the first membrane receptor uptake mechanism to be fully described (Goldstein and Brown; 1977). This achievement by Brown and Goldstein was rewarded by the Nobel Prize Committee.

Redgrave and coworkers (1992) have examined the metabolism of emulsion droplets resembling chylomicrons (*cf.* Marankao *et al.*, 1986). The emulsions were prepared without any protein by using phospholipids only for emulsification. When injected into the bloodstream of rats, these particles appeared to be metabolized like chylomicrons. A series of interesting features was observed in the degradation of different lipid classes. The presence of cholesterol is necessary in order to mimic the metabolism of lipoproteins. Free cholesterol seems to regulate the metabolism by influencing the binding of apolipoproteins to the surface of the emulsion droplet. Thus, free cholesterol favours the binding of apoprotein E to the emulsion droplet surface, whereas the binding of apoproteins A-I and A-IV is reduced.

C. LIPID-PROTEIN INTERACTION IN LIQUID CRYSTALLINE PHASES

Extensive studies of the liquid-crystalline phases formed in various ternary systems of lipids, proteins and water led to the conclusion that aqueous lipid-protein liquid-crystalline phases are only formed when there is an electrostatic interaction between the lipid and the protein (Gulik *et al.*, 1969; Rand, 1971). Later studies have shown that cubic phases of non-ionic lipids can accommodate proteins, however (Ericsson *et al.*, 1983). It is therefore natural to discuss these two types of interaction separately.

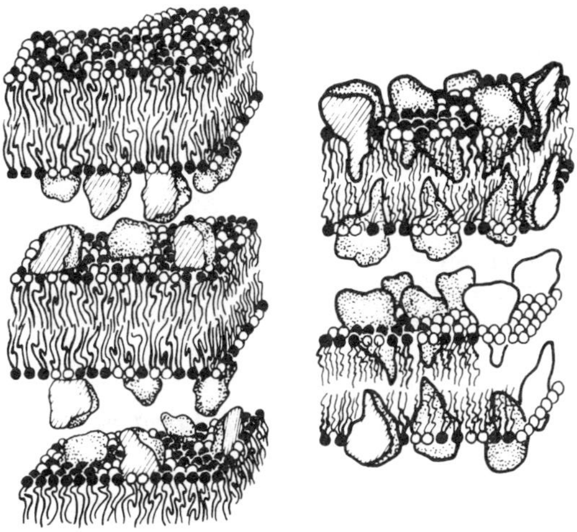

Fig. 6.1. The two alternatives of protein incorporation in the L_α-phase. The ionic association with the whole protein in the aqueous medium is shown to the left; the association involving penetration of the protein into the hydrocarbon core of the bilayer is shown to the right, after Rand (1973).

1. Interaction between Lipids and Proteins with Opposite Net-Charge in the L_α-phase

If a dispersion of the L_α-phase (its structure is discussed in Chapter 7) of non-ionic polar lipids is exposed to water-soluble proteins, there is no sign of molecular interaction. On the other hand, the introduction of only a minor proportion of a charged lipid results in phase separation when exposed to proteins of opposite net charge. These observations formed a logical basis for concluding that lipid-protein interactions "are unlikely and exceptional" in cases where there is no electrostatic association (Rand, 1973).

The formation of a lipid-protein-water L_α-phase can take place in two ways, and the corresponding structures have been derived from their X-ray diffraction data (Rand, 1971). The two alternatives are indicated in Figure 6.1.

The interaction between insulin and a PC/cardiolipin L_α-phase was found to give an electrostatic complex at a pH where the protein has a positive net-charge. In this case, the protein is located in the water layer. The other alternative was observed when a PC/cardiolipin L_α-phase was exposed to bovine serum albumin at a pH where the net-charge of the protein was opposite to that of cardiolipin. The X-ray data showed that the protein in this case is partly buried in the lipid bilayer, as shown in Figure 6.1.

The incorporation of membrane proteins into the L_α-phase is one way of reconstituting cell membrane structures. These proteins are part of the hydrocarbon core, sometimes spanning the bilayer. The structure of the bilayer is thus similar to that of the model system shown in Figure 6.1.

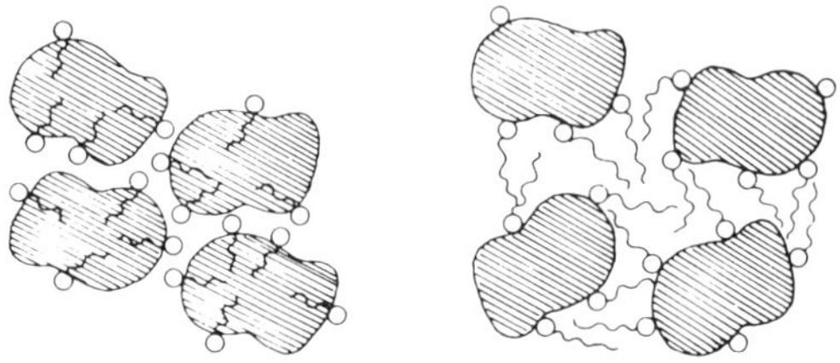

Fig. 6.2. Lipid-protein complex of insulin and phosphate esters. At a chain-length of C_{10}, the hydrocarbon chains are buried in the protein structure, as shown to the left. With longer chains (C_{16}) or cardiolipin , the hydrocarbon chains forms a continuum, after Larsson and Rand (1973).

2. Aqueous Precipitates of Amorphous Lipid-Protein

Electrostatic interaction between lipids dispersed as an L_α-phase (see Chapter 7) and proteins can also give amorphous precipitates. They do not even show a 4.5 Å band in the X-ray diffraction pattern, a finding which proves that there are no continuous hydrocarbon chain regions, as determined by studies on insulin and phosphate esters or cardiolipin (Larsson and Rand, 1973). According to a Raman spectroscopy analysis, the protein remain in its native conformation, and there are two alternative structures of the complex as shown in Figure 6.2.

When all lipid molecules are ionic with a charge opposite to the net-charge of the protein, there seems to be a general tendency to formation of amorphous lipid-protein-water phases. One reason might be that the amount of lipid in these complexes is too small to form bilayer types of structures.

3. Cubic Lipid-Protein-Water Phases

When it was realised that the cubic lipid-water phase consists of open water-channel systems separated by the infinite lipid bilayer, it was natural to examine whether this phase could accommodate proteins. The ternary system lysozyme-monoolein-water was therefore studied, in which monoolein was chosen in order to have a non-ionic lipid (Ericsson *et al.*, 1983). The phase diagram is shown in Figure 6.3.

The most remarkable feature was the observation that large amounts of protein can be incorporated to form lipid-protein-water phases without an ionic interaction, contrary to what was known from earlier work. By examination of the denaturation enthalpy of the protein (from differential scanning calorimetry (DSC) data), it was further shown that all of the protein molecule were located in the water region in their native conformation. From the appearance of the phase boundaries (Figure 6.1), it can be seen that the presence of the protein swells the cubic phase (as is also evident from the unit cell size). The cubic region in this system shows all the three fundamental IPMS-structures described in Chapter 3 in fact,

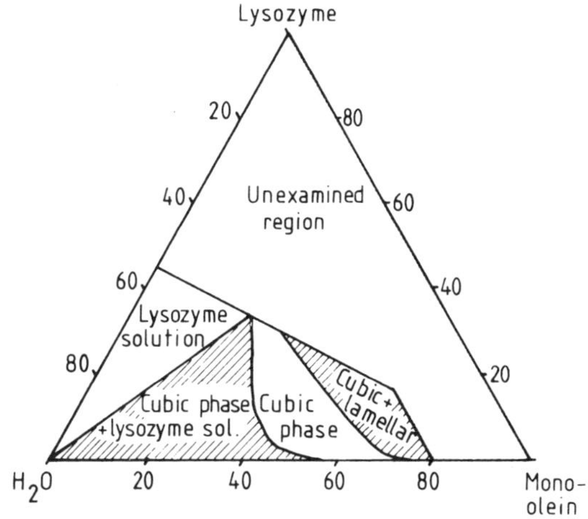

Fig. 6.3. Ternary phase diagram of monoolein/water/lysozyme.

i.e. the gyroid surface C_G, the diamond surface C_D and the primitive surface C_P in the order of increasing protein/water ratio (Buchheim and Larsson, 1987).

The cubic phase can accommodate various proteins of sizes at least up to about 150,000 Daltons (Ericsson *et al.*, 1983). A study of the electron freeze-fracture microscopy texture of the cubic phase of sunflower oil monoglycerides/water/cytochrome c (Gulik-Krzywichi *et al.*, 1984) has been reported. An electron micrograph is shown in Plate 6.1.

The body-centred arrangement in the C_P-structure can be observed. The texture also illustrates the single-crystal character of cubic phases. It could finally be mentioned that the incorporation of a hydrophobic protein, wheat gliadin, into the cubic monoolein-water phase has been studied (Larsson and Lindblom, 1982). The cubic phase can contain a protein/lipid ratio of up to 3:7. It should be expected that this protein is partly located in the hydrocarbon chain region of the lipid bilayer.

D LIPID-PROTEIN INTERACTION IN MONOLAYERS

1. Adsorbed Monolayers

When the lipid (surfactant) as well as the protein are water-soluble, both will form adsorbed monolayers at the air/water interface. If the surface tension is recorded as a function of concentration of the components, it is possible to determine if there is a molecular interaction or not. The two alternatives are illustrated in Figure 6.4. An example of a mixture with no interaction is monocaproin and ovalbumin (Ericsson and Hegg, 1985).

The surface tension curve of the pure surfactant shows the characteristic steep reduction at the CMC-value. When a protein is present, it gives a

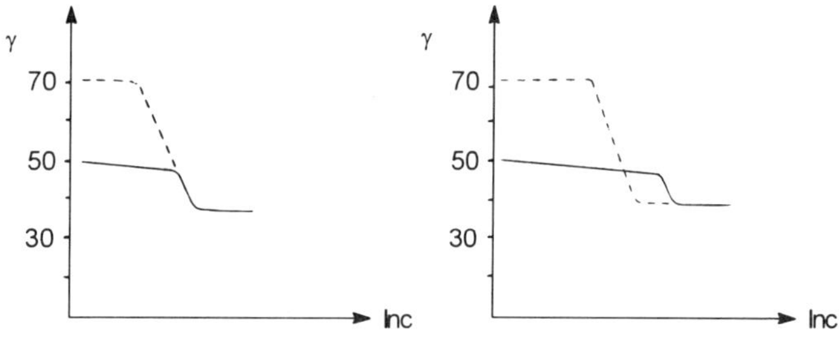

Fig. 6.4. Surface tension behaviour of adsorbed mixtures of proteins and surfactants, showing no interaction (to the left) and molecular interaction (to the right). The broken curve shows the surface tension versus surfactant concentration without any protein present. The continuous line shows the surface tension versus surfactant concentration at a constant concentration of protein.

certain reduction in surface tension because of the film of adsorbed protein at the air/water interface. When the surfactant is added to this protein solution in the case of no interaction, the film is not influenced until a surfactant concentration is reached where the reduction of surface tension of the surfactant approaches that of the protein solution. At higher concentrations, the surfactant film will squeeze out the protein molecules from the interface. The fact that the mixed film follows the curve of the pure surfactant at high concentrations shows that there is no interaction.

When the two components interact, however, there is a shift between the curve of the mixed film and that of the pure surfactant at high concentrations. The reason is that the protein and the surfactant form a complex in the solution. The change in the complex composition with surfactant concentration can be seen in the change in surface tension versus concentration. Thus the adsorbed monolayer consists of a surfactant-protein complex (otherwise the surface tension would be constant with respect to concentration). On the other hand, at a certain surfactant concentration, the surfactant will also squeeze out all of the lipid-protein complex from the interface. The position of the steep reduction of surface tension then obtained is an indication of the amount of surfactant in the surfactant-protein complex. This type of behaviour is usually due to electrostatic interaction, and has been demonstrated in mixtures of ovalbumin and SDS, for example (Ericsson and Hegg, 1985).

2. *Protein Interaction with Insoluble Lipid Monolayers*

The monolayer technique discussed in Chapter 5 can be used to study lipid-protein interactions in monolayers. If the protein is very hydrophobic, as with cereal storage proteins, they can be dissolved in an organic solvent. The lipid and protein are then dissolved and spread together on the water surface. However, the protein cannot usually be spread from solvents, and the lipid monolayer is then spread first. During compression with recording of the Π-A isotherm, the protein can be injected into the substrate at different positions along the Π-A isotherm.

In most cases without an ionic interaction in bulk, there is no molecular interaction between the lipid and the protein in the monolayer. At a low pressure, below about 20 mN/m, a mixed film is formed, consisting of patches of one phase dispersed in the other. The lipid-soluble fluorescent probes described in Chapter 5 can be used to determine if the two components segregate or form molecularly mixed films. At higher pressures (in the region 30-40 mN/m), the protein molecules are usually squeezed out from a monolayer of polar lipids. Triglycerides, in contrast, show such low collapse pressures that the triglyceride molecules can be squeezed out from a mixed film formed at low pressures, resulting in a pure protein monolayer.

There are, however, certain proteins which can penetrate lipid monolayers at very high film pressures. Bacterial phospholipase C can cleave phospholipid molecules when injected under the water phase of phospholipid monolayers at pressures above 40 mN/m, and the reaction requires penetration of the monolayer (Titball, 1993). In eukaryotic cells, phospholipase C has an important role in metabolic regulation by the generation of second messengers, diacylglycerols and inositol triphosphates, formed by this cleavage reaction.

The formation of prostaglandins, leukotrienes and related hormones is controlled by the release of arachidonic acid, and phospholipase A_2 cleaves phospholipid molecules at the 2-position where this fatty acid is located (*cf.* Dennis, 1987). Phospholipase A_2 acts on the monolayer conformation of PL molecules, and when one PL molecule is bound to an "activator" site, another molecule is bound to the catalytic site where cleavage takes place (Dennis, 1987).

When there is a molecular interaction in this type of monolayer, there are two extremes, which are analogous to the interaction in bilayers shown in Figure 6.1. Thus an electrostatic interaction between an ionic lipid and a protein with the opposite net-charge can result in an association of the protein to the lipid monolayer, with the whole protein located in the water phase . The lipid monolayer can also be penetrated by hydrophobic regions of the protein. This alternative corresponds to a stronger interaction, and represents the general solubilization mechanism of membrane proteins into lipid mono- or bilayers.

E. LIPID INTERACTION WITH OTHER POLYMERS

There are two main reasons for the interaction between lipids and other polymers on a molecular level. If the polymer is surface-active, the situation is the same as in proteins as discussed above. Interactions in the bulk as well as at interfaces should therefore be expected. Alternatively,

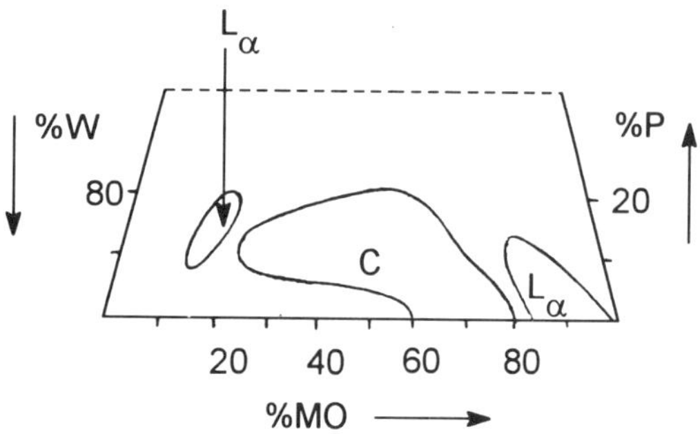

Fig 6.5. Part of the phase diagram of a pluronic/water/monoolein system, after Landh (1994). Only the bottom part of the ternary system is illustrated, with one-phase regions indicated. The cubic region is denoted "C".

there can be specific interactions resulting in an energetically more favourable conformation of a polymer when it forms a lipid complex. A good example is the interaction between amylose and non-ionic lipids. The helical inclusion complex formed will be discussed in detail in connection with starch in Chapter 14. We will only consider the first of these mechanisms behind lipid-polymer interactions below.

Pluronics are non-ionic surface-active block-polymers, consisting of segments of poly-ethyleneoxide alternating with those of poly-propyleneoxide. This polymer can interact with polar lipids on a molecular level. A complete phase diagram involving monoolein has been reported, and the main features are shown in Figure 6.5 (Landh, 1994). The most interesting feature of this diagram is the complex cubic region, exhibiting many different cubic phases. Furthermore, an L_α-phase is formed at a high water content. The three-phase region L_α/cubic phase/ water was found to be ideal for the generation of dispersed particles of the cubic phase discussed in Chapter 7.

Another polymer interacting with polar lipids, which has been thoroughly investigated, is ethyl-hydroxyethyl-cellulose (EHEC) (*cf.* Carlsson *et al.*, 1990). Among the strongly polar lipids (surfactants) studied with regard to interaction with the EHEC-molecule, the ionic ones were found to give a remarkable gelling phenomenon. This gel is formed reversibly on heating, and the gelling temperature can be controlled by variation of the polymer structure and the type of surfactant. The structure of the complex in the gel is considered to involve micellar aggregates of the surfactant at hydrophobic regions along the polymer chain. These micelles are proposed to cross-link different polymer chains, in this way forming a three-dimensional network.

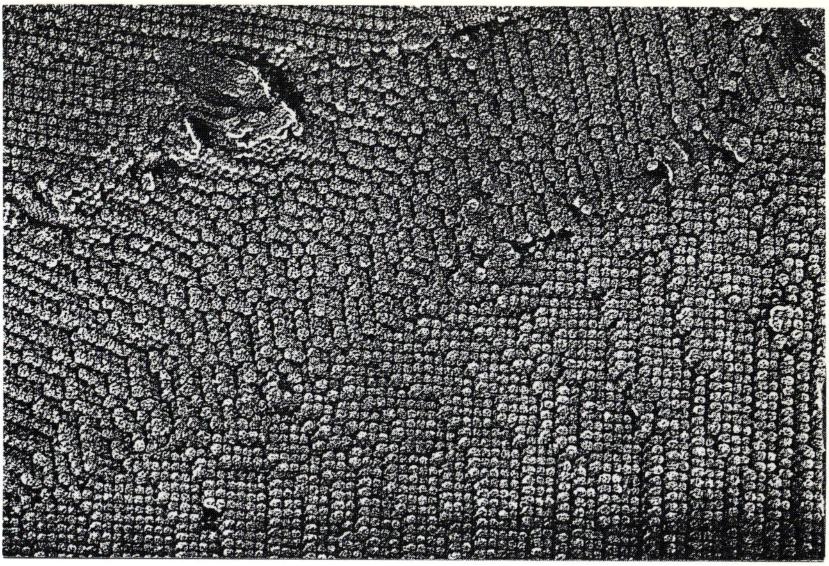

Plate 6.1. Electron micrograph of the cubic phase in the system sunflower oil monoglycerides/water/cytochrome C, after Gulik-Krzywichi *et al.* (1984).

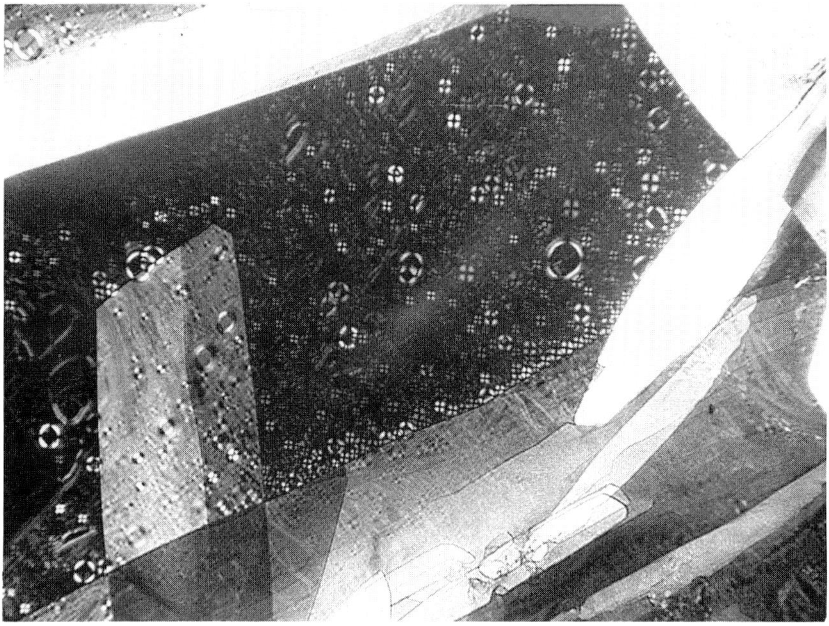

Plate 7.1. Crystallization of monocaprin from a liposomal dispersion as seen in the polarizing microscope, after Larsson (1968).

CHAPTER 7

DISPERSIONS OF LIPID-WATER PHASES

What happens when two or more phases coexist in lipid-water systems? Can they be dispersed in one another, and what kind of flow properties are obtained in such dispersions? These questions are of technical significance in many applications. One special case is emulsions, where oil and water phases coexist. Emulsions will be treated in Chapter 8, whereas the dispersion properties of aqueous lipid phases are considered in this chapter. Because water is the continuous phase in most applications as well as in biological tissues, this chapter will be focused on aqueous dispersions of the various lipid-water phases described in Chapters 3 and 4.

A. DISPERSION OF THE L_α-PHASE - LIPOSOMES AND VESICLES

In the early 1960s, Bangham showed that phospholipids in diluted aqueous systems form aggregates consisting of concentric lipid bilayers. It was later shown that the general condition for existence of such spherically (or cylindrically) closed lipid bilayer structures is the occurrence of a two-phase region, *i.e.* that the lamellar liquid-crystalline phase coexists with water (Larsson, 1967b). Stable dispersions were obtained by mild flow only, and the rheological properties of this region of the phase diagram were therefore examined (Krog and Larsson, 1968). As the L_α-phase was dispersed almost spontaneously, it seemed natural to term this region "dispersion". Most polar lipids, especially membrane lipids, exhibit this two-phase region (L_α-phase + H_2O). These dispersions were later called liposomal dispersions. It was soon realised that this type of aggregate, closely resembling biomembranes, might be used for encapsulation of drugs and drug delivery. The technology of liposomes has been developed mainly by Papahadjopoulos (*cf.* Szoka and Papahadjopoulos, 1980) and by Gregoriadis (1990), and it will be discussed in Chapter 13.

The term liposome is sometimes used to describe single-walled closed

Fig. 7.1 A possible mechanism behind formation of a liposome from the L_α-phase.

bilayers also, but it is more usual to term them *vesicles*. We will use the latter term here, and the term *liposome* is reserved for spherically concentric *multi-bilayer* aggregates. The curvature gives a limitation to the size of the smallest possible vesicle, and by experience this is known to be about 200-300 Å in diameter (to be consistent with a phase diagram requiring equilibrium with the lamellar liquid crystalline phase). Vesicles can range in size up to the μm-range ("giant" vesicles), however.

From what we have learned about the lamellar liquid-crystalline phase (Chapter 3), it is evident that well-defined liposomes can be prepared only when we have limited swelling of these phases. With non-ionic or zwitter ionic lipids, the thickness of the water layer of about 20 Å provides good conditions to form liposomes. When we have charged lipids present, however (even the presence of a few percent charged lipid among non-ionic lipids is enough), their "infinite" swelling behaviour means that they are not suited for formation of multilayer aggregates. They will instead give quite stable vesicle dispersions, because of the electrostatic repulsion. Non-charged lipids can of course also be used to prepare vesicles. The presence of charges on the surface is an important feature in targeting. Positive charges, for example, will tend to induce fusion with cells having a negative net-charge. The mechanical properties of the lipid bilayer, defined by so-called fluidity as measured by molecular mobility, will also influence the stability of liposomes and vesicles. Cholesterol is a lipid component often used to stabilize liposomes.

The main reason for the limited applications of liposomes so far is the fact that they are not thermodynamically stable. They resemble emulsion droplets in this respect, possessing only kinetic stability. Whether or not vesicles can be thermodynamically stable under special conditions is a question under debate at present, and we will not enter this discussion here.

Much work has been devoted to liposome technology in order to produce monodisperse and stable preparations. In principle, liposomes can be prepared by any mechanical disruption of the lamellar liquid crystalline phase. The reason is simple. When lipid bilayers separated by water are fractured, they will tend to fuse to avoid contact between hydrocarbon chains and water. There are many other possible deformations giving the closed liposomal structure on fragmentation of the L_α-phase. A hypothetical situation when liposomes will form is indicated in Figure 7.1. This kind of flow can be visualized in the polarizing microscope.

A commonly used mechanical method for the production of liposomes and vesicles is ultrasonification, but a disadvantage is the induction of lipid peroxidation. By using certain buffers, it is possible to reduce such lipid degradation (Fiorentini *et al.*, 1989).

The most important application of liposomes is for drug delivery, and the technological aspects of liposomes, for example permeability, are therefore discussed in Chapter 13. It should finally be pointed out that the literature on liposomes is expanding very rapidly. There is even a specialized journal in this field, *Journal of Liposomal Research* (Marcel Dekker, New York).

B. DISPERSION OF REVERSED HEXAGONAL AND OF CUBIC PHASES - CUBOSOMES

If a reversed hexagonal phase or a cubic phase in excess of water is deformed by some mechanical technique, it is not possible to disperse the phase. When two pieces of such a phase come into contact in water, they fuse immediately. There is, however, still a possibility of forming a dispersion of these phases. We will first consider the cubic phase.

If the cubic phase is broken up into pieces in a liposomal dispersion, small particles can be obtained with an L_α-bilayer surface, and a core consisting of the cubic phase (Larsson, 1989). It was natural to term these particles *cubosomes* by analogy with liposomes. The structure is shown in Figure 7.2.

If we consider fractures through a cubic phase in water, *e.g.* the one shown in Figure 3.8, it will break preferably through the "necks", where the Gaussian curvature has its highest negative value. It will then expose the hydrocarbon chain core towards water, unless it can fuse with another lipid bilayer. If there is an L_α-bilayer present, fusion can result in an L_α-bilayer with "holes" towards an inner cubic core. It is also possible that such cubic particles can have one or more "ordinary" L_α-bilayers on the surface. The structures of cubosomes have been examined by X-ray diffraction and electron microscopy (Landh, 1994), and the structure described above has been confirmed. By use of a valve homogenizer or by ultrasonification it is possible to produce monodisperse cubosomes in the range 0.1 - 10μm. (Landh, 1994). An efficient method of producing cubosomes is to use surface-active polymers. It seems probable that there is then no outer L_α-bilayer, but rather a protective-colloid type of coverage of the exposed hydrocarbon chain cores of fractured cubic planes. The ternary system of monoolein/water/pluronic shown in Figure 6.5 appeared to be an ideal system for cubosome production (Landh, 1994). Another system which easily gives cubosomes is monoolein/caseinates/water (Buchheim and Larsson, 1987).

Cubosomes have been shown to occur frequently in biomembrane assemblies, and this topic is discussed in Chapter 9. Furthermore, they are

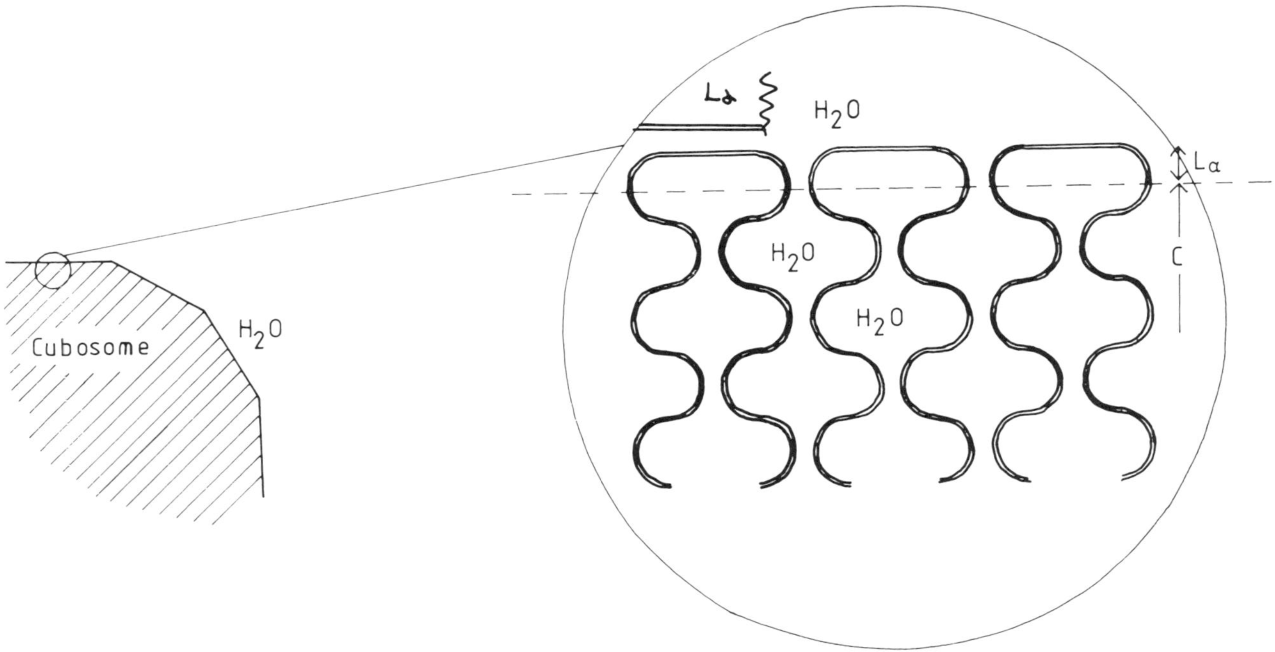

Fig. 7.2. Schematic illustration of a cubosome. A cross-section through a surface zone is shown above (the lipid bilayer is drawn). There might be additional bilayers of L_α-conformation outside as indicated on the left side of the detailed illustration.

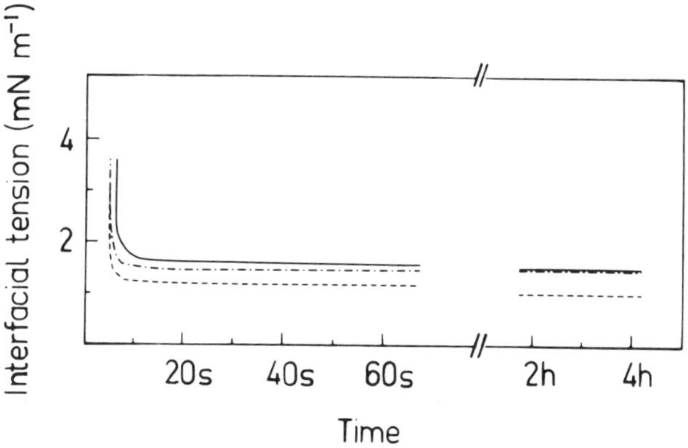

Fig. 7.3. Interfacial tension according to the drop/volume-technique of the L2-phase against water (continuous line), against 0.2 M sodium chloride (alternating short and long lines), and against a caseinate solution (1% w/w in 0.2M sodium chloride) (dotted curve), after Pilman *et al.* (1982).

discussed in connection with drug delivery in Chapter 13. The technique of dispersion of cubic phases by surface-active polymers can be applied in the same way in order to disperse the H_{II}-phase.

C. DISPERSION OF L2- AND L3-PHASES

The dispersion of a liquid phase in another liquid phase is by definition an emulsion. As the L2- and L3-phases are themselves thermodynamically stable aqueous phases, such emulsions are in fact a kind of water-in-oil-in-water double emulsion. We will first consider the dispersion of the L2-phase, using the triglyceride-based L2-phase defined in Figure 3.26. The interfacial tension between an L2-phase and water is shown in Figure 7.3.

The interfacial tension is very low, less than 2 mN/m. The corresponding values for a triglyceride oil and a mineral oil are about 30 and 50 mN/m respectively. The presence of proteins in the water solution gives an additional lowering of the interfacial tension. Proteins were introduced to stabilize emulsions of the L2-phase in water (Pilman *et al.*, 1982).

By mechanical homogenization of the L2-phase in excess of water, a quite stable emulsion is obtained. It can be further stabilized if the water phase contains a protein. By using very surface-active proteins, like caseins or caseinate, a well-defined emulsion can be prepared. Such preparations are water/oil/water emulsions. It is not obvious where the protein molecules are located in such emulsified particles. In the case of globular proteins, it is easy to follow how coloured ones (like myoglobin and cytochrome c) behave. It can then be seen that such proteins are solubilized in the water aggregates of the L2-phase. This diffusion process

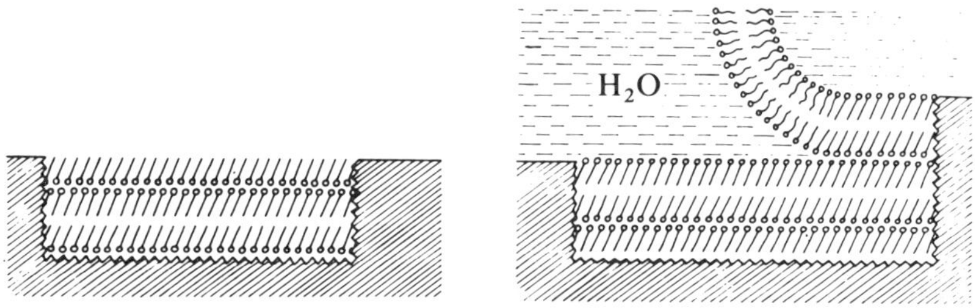

Fig. 7.4. Crystallization from the liposomal dispersion, resulting in exposure of polar head groups on the dominating crystal surfaces. A fragment of the corresponding crystal grown in an organic solvent is shown above. Its surface formed by methyl end groups make this crystal hydrophobic, after Larsson (1968).
Legend to Plate

is very slow, however. By emulsifying an L2-phase directly after it has been exposed to the protein solution, the protein molecules will be adsorbed at the L2/water interface, as evident from Figure 7.3. After the emulsion has been formed, there will then be a transport of protein molecules into the core of the L2-droplets. At equilibrium, three different protein concentrations should be expected, one within the L2-droplets, another at the interface and a third in the external water phase.

What has been described above in dispersion properties of the L2-phase is also valid in the case of L3-phases. The main difference is that such aqueous dispersions contain much more water, as L3-phases occur in the water-rich region of the phase diagram.

D. AQUEOUS DISPERSIONS OF HYDROPHILIC LIPID CRYSTALS

In connection with studies of dispersions of the L_α-phase of monoglycerides, it was observed that crystallization from this two-phase region gave dispersions of lipid crystals with a remarkable "water-binding" capacity (Larsson, 1968). In the polarizing microscope, it was possible to follow the transformation of lipid bilayers in liposomes to crystal bilayers. This growth is illustrated in Plate 7.1.

The proposed mechanism of the crystallization process is indicated in Figure 7.4. "Ordinary" monoglyceride crystals, grown in organic solvents or from the melt, have hydrocarbon chain methyl end groups forming the dominating surfaces of the platelet-shaped crystals. Crystallization from the liposomal dispersion, on the other hand, seems to involve a direct transformation of lipid bilayers from the L_α-conformation to the crystalline conformation. The dominating surfaces will therefore always consist of the polar head groups, which means that the crystals become hydrophilic.

The density of the hydroxyl groups on the dominating crystal surface is evident from the crystal structure (Chapter 2), and it is as high as one group per 11 Å^2. This makes the crystals very hydrophilic. Up to 90% (w/w) of water can thus be associated to the crystal surfaces, giving a semi-solid mass of the lipid-water mixture, that appears homogeneous when viewed by the naked eye. This phenomenon has applications in foods and in drug delivery, as will be discussed on Part 2.

Part 2

Lipid Functionality

CHAPTER 8

EMULSION SCIENCE AND TECHNOLOGY

The word emulsion originates from the Latin, *emulgeo*, meaning to milk. There are numerous applications of dispersions of mineral or triglyceride oils in water, *i.e.* oil-in-water (o/w) emulsions. Less common, but still of technical importance, is the reversed type, with an aqueous solution dispersed in an oil phase, *i.e.* water-in-oil (w/o) emulsions. Emulsions are thermodynamically unstable, and the key issue of emulsion technology is dispersion stabilization in order to get acceptable shelf-life. Emulsion stability is therefore a question of kinetic stability. There are also more complex emulsion types, like *double-emulsions* (o/w/o or w/o/w), which were discussed in connection with L2- and L3-dispersions (Chapter 7). If we emulsify more than two liquids, which do not mix, *multiphase emulsions* can be formed. Silicon oil, for example, could be dispersed in a vegetable oil, which in its turn is dispersed in water.

We will only consider "ordinary" o/w- or w/o-emulsions in this chapter. Triglyceride oils are the most important type of lipid oil from a technological point of view, but the discussion and conclusions below are also relevant to emulsions based on mineral oil. So-called microemulsions were discussed in Chapter 7. "Ordinary" emulsions are sometimes termed macroemulsions, not to be confused with microemulsions. The droplet size in a macroemulsion is larger than about 1 μm in diameter.

A true description of emulsions has evolved as recently as during the last two decades, and emulsion technology is still often based more on trial-and-error than on a scientific understanding of the actual system.

A. FORCES BEHIND EMULSION STABILIZATION

1. A Model Emulsion

A simple system that can form quite stable emulsions can be obtained by dispersing phospholipids into a triglyceride oil and then emulsifying the oil in water. Emulsions for parenteral nutrition are prepared in this way, often with egg yolk or soybean oil phospholipids as emulsifier. In the preparation of the emulsifier from the crude lipid raw material,

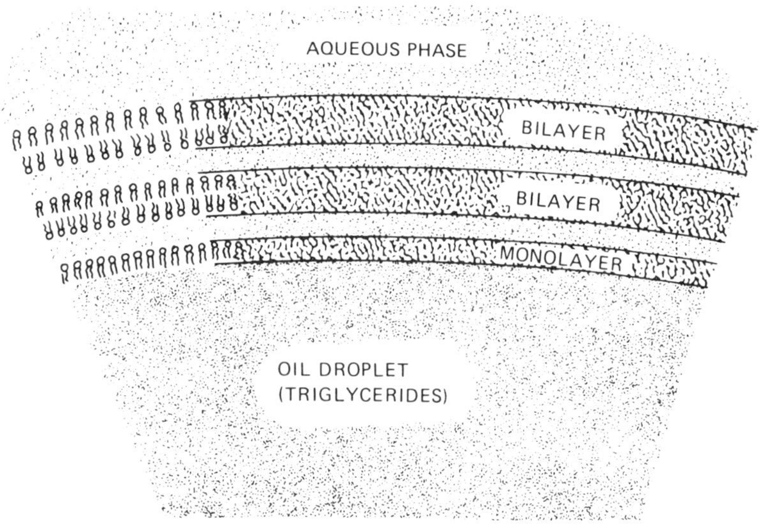

Fig. 8.1. Structure of an emulsion prepared from phospholipids and a triglyceride oil.

predominantly phosphatidylcholine, a few percent of anionic species left in the final mixture will improve the efficiency of the emulsifier. The presence of the charges will contribute to the electrostatic repulsion, as discussed below. It is also possible to emulsify the oil with pure PC. From our knowledge of the interaction of different types of lipids in aqueous systems and phase equilibria described in Part 1, it is possible to describe the arrangement of the different molecules in such an emulsion, as shown in Figure 8.1.

From the Π-A isotherm of mixtures of triglycerides and phospholipids at the air/water interface, it can be seen that almost all of the triglyceride molecules are squeezed out from the monolayer at pressures above 20 mN/m. The solubility of phospholipid in water is about 10^{-11} M, and in the oil it is less than about 0.1% (w/w). An emulsion having a higher level of phospholipids will therefore give a monolayer pressure equal to the equilibrium spreading pressure, and it will exist in equilibrium with the bulk phase of the phospholipid, which is the L_α-phase. From this it can also be concluded that the phospholipids form a monolayer with the same conformation as in the L_α-phase between the triglyceride core and the outer water phase. It should be mentioned that it is even possible to perform direct measurements of phospholipid monolayer behaviour according to Π-A isotherms at the oil/water interface. From an experimental point of view, such studies are complicated, however. This is the reason for considering the air/water interface as a model of the oil/ water interface in emulsions.

2. Emulsion Breakdown Processes

The stability of an emulsion is limited by the following processes:

- *Flocculation* of emulsion droplets, which is an aggregation phenomenon with the droplet interface intact and a liquid film in between aggregating droplets. Flocculation leads to an increased viscosity, and sometimes even formation of a gel network.
- *Coalescence* of flocculated droplets. This fusion process results in a reduction in the viscosity. When coalescence has proceeded in many steps, the droplet density difference in relation to the continuous phase will lead to the formation of an oil-rich layer at the surface. In connection with o/w-emulsions, this phenomenon is termed *creaming.*

These processes will be considered in more detail in discussions of emulsion destabilization at the end of this chapter.

3. The DLVO-Theory of Colloidal Stability

About half a century ago, the following simple and very useful theory for colloidal dispersions was derived by Deryaguin and Landau in the Soviet Union and Verwey and Overbeck in Holland. After them, the theory is known as the DLVO-theory. There are two main requirement for a colloidal dispersion to be kinetically stable over a long time. The first is that there is no sedimentation. This means that the particle diffusion rate must exceed the sedimentation rate. The second factor is that the particles do not aggregate, and these conditions are defined by the DLVO-theory.

If we consider the surface of an emulsion droplet, such as that of our model in Figure 8.1, negative charges on the droplet surface will result in adsorbed cations. The adsorbed counter-ions are hydrated and mobile, reducing the surface potential with the distance *d* from the surface by:

$$e^{-\kappa d}$$

where κ is the Debye screening length, which is proportional to the counterion charge and the square root of the salt concentration of the water phase.

This electric double-layer will give an electrostatic repulsion between the particles. The repulsion falls off with the distance between the particles according to:

$$e^{-d/\kappa}/d$$

There is also an attractive force between the particles; the van der Waals attraction. At particle distances well above the particle size, this attraction is equal to:

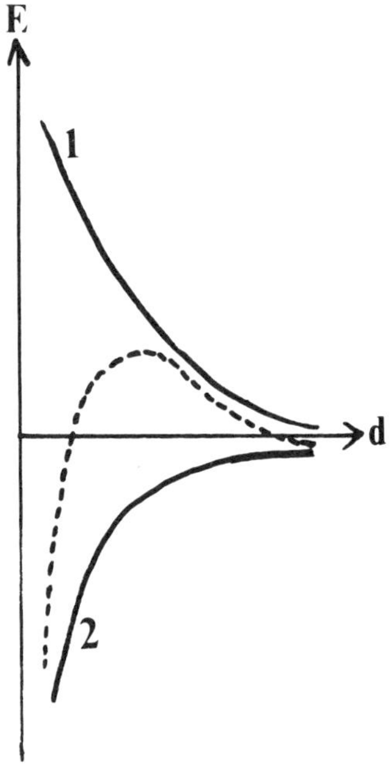

Fig. 8.2. Interaction energy between two particles: dotted curve, equal to the sum of the electrostatic repulsion (curve 1) and the van der Waals attraction (curve 2).

$$H / d^6$$

where H is the Hamaker constant.

The interaction energy between two particles is the sum of the electrostatic repulsion and the van der Waals attraction. These two curves and their sum are illustrated in Figure 8.2. When the particles approach each other, they have to overcome a barrier in order to come into contact. This barrier is the reason for the kinetic stability of emulsions. Just before this barrier, there is a weak minimum in the interaction curve, which sometimes results in flocculation.

The DLVO-theory has been applied successfully in emulsion studies for a long time. The theory has been tested by direct force measurements using the surface-force apparatus (Israelachvili, 1985), and good agreement with the theory at long distances has been observed. At short distances, the hydration force described in Chapter 3 must also be taken into account.

4. Steric Stabilization

Imagine that a glycolipid with many sugar residues is solubilized in the interfacial monolayer of the polar lipids shown in Figure 8.1. Such hydrated chain tails surrounding emulsion particles will hinder them from coming into contact, and provide a stabilization mechanism that is called steric stabilization. Polymers are also effective steric stabilizers of emulsions, due to adsorption on the surface with loops and end tails forming a barrier against flocculation and coalescence.

The stabilization of emulsions by casein fractions is due to a combination of steric stabilization with electrostatic repulsion, as will be considered in Section B.

5. *Polar Lipid Phases at the Oil/Water Interface*

The ability of polar lipid crystals or aqueous phases of lipids to form both hydrophobic and hydrophilic surfaces is the mechanism behind emulsion stabilization based on polar lipids.

Friberg and co-workers were first to observe that the three-phase region oil/water/L_α-phase gives extremely stable emulsions in ternary systems involving non-ionic surfactants (Friberg *et al.*, 1969).

The model emulsion shown in Figure 8.1 has an outer phospholipid-water shell. As pure PC was used in the emulsification, the interface consists of an L_α-phase. The requirement for this is that there is enough PC present in relation to the total area of the oil/water interface.

In Section B, we will also consider other such emulsions, with an interface consisting of a separate phase of polar lipids, which forms a hydrophobic surface towards the oil and a hydrophilic phase towards water.

6. *Viscosity of the Continuous Phase*

It is obvious that an increased viscosity of the continuous phase will result in reduced mobility of the dispersed droplets, and this means reduced flocculation and coalescence. Sausage is sometimes described as an emulsion. It can also be described as an aqueous gel, where oil droplets and fat cells are dispersed. The emulsion description is adequate before heat treatment, as the consistency of the continuous phase is liquid. Heat treatment results in denaturation of myofibrillar proteins and collagen. After cooling the emulsion droplets are therefore immobilized by the gel structure of the aqueous medium. This is also the dominating stabilization mechanism of w/o-emulsions; a network of fat crystals immobilizes the oil phase.

B. OIL-IN-WATER EMULSIONS

1. *Emulsification*

Dispersion of one liquid phase into another requires energy input. The minimum energy needed is:

$$A\gamma$$

where A is the area of the droplets (or rather the increase in the interfacial area) and γ is the interfacial energy (which can be measured as interfacial tension). The energy can be supplied by a stirrer, for example by an Ultra-Turrax. The higher the shear rate, the smaller are the droplets formed. The mechanical principles behind emulsion formation have been extensively studied by Walstra (1993). A droplet in a laminar flow will burst when the velocity gradient exceeds a critical value. In cases of a continuous phase with low viscosity, like water, and a dispersed phase with high viscosity, very high velocity gradients are needed. More effective is elongational flow, with no shear and a velocity gradient in the flow direction. Even viscous drops are then deformed and are broken into smaller ones. Turbulent flow is much more effective than laminar flow in breaking up droplets, when the continuous phase has a low viscosity. The local flow velocity then varies in a chaotic way. The disadvantage is a spread in the particle size distribution, however.

The average droplet diameter using the same energy input is strongly reduced when going from an Ultra-Turrax to a high-pressure valve homogenizer. Ultrasonification gives a particle size distribution in between. A log/log plot of the droplet diameter versus energy input shows a linear relationship (Walstra, 1993). When using valve homogenization or ultrasonification, an important mechanism is the formation of cavities by pressure fluctuations. In the valve homogenizer, the oil-water mixture is forced through the valve opening, and this gives a steep pressure gradient. Milk is homogenized in this way. By decreasing the particle size of the native milk fat globule, flocculation and the creaming tendency are eliminated. If the viscosity of one or both liquids is very high, it is better to use a colloid milk. The two liquids are then deformed together through a thin slit. On ultrasonification, the dispersion takes place via the cavities thus formed. An ultrasound intensity of about 3 kW per square meter will give fluctuations of 1 atmosphere.

2. *Protein-Stabilized Emulsions*

A protein will orient and usually change its conformation at an interface, in order to reduce the interfacial energy. The native conformations of protein molecules, such as the α-helix and β-sheet structure, are stabilized by hydrophobic interactions between non-polar amino acid residues. The presence of an oil-water interface would be expected to induce a reconformation of the peptide chain, with the non-polar groups facing the oil. In the earlier literature on proteins at interfaces, it was usually assumed that the peptide chain was unfolded. The present picture is somewhat modified, however. Based on ellipsometry and surface-force studies of proteins at interfaces, it has been realised that the molecules tend to be globular and compact in the adsorbed state also (*cf.* Wahlgren and Arnebrant, 1991).

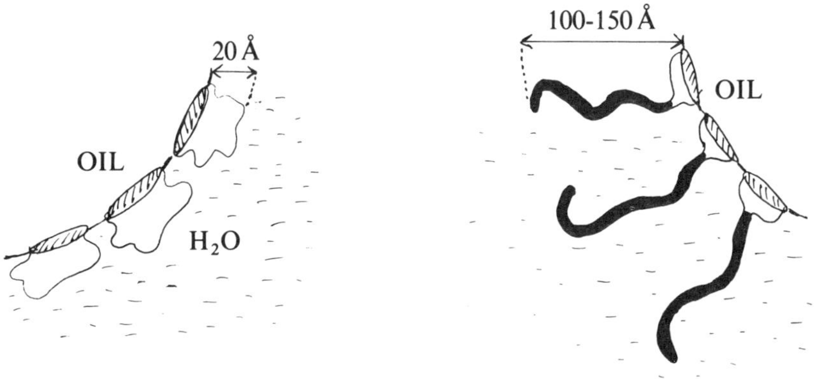

Fig. 8.3. Schematic illustration of the arrangement of globular proteins such as β-lactoglobulin (to the left) and β-casein (to the right) at the interface of o/w-emulsions.

There are two main types of protein adsorption behaviour at the oil-water interface. Proteins with well-defined globular tertiary structures, like whey proteins, adsorb with a conformation close to that in the native state; whereas proteins which lack a well-defined tertiary structure are more unfolded after adsorption, with hydrophilic segments extending into the water phase. These two cases are illustrated in Figure 8.3. β-Casein is a good example of the second type, with residues 1 to 48 located in the water medium (Dickenson, 1992). Because of the presence of many negative charges, this segment is quite extended (100-150 Å), providing an effective steric stabilization. The casein fractions are extremely effective in stabilizing emulsions.

The viscosity of the interfacial protein film provides information on its structure. The shear viscosity of films of close-packed globular proteins is much higher than that of flexible casein monolayers at the interface (Dickenson, 1992). Interesting ageing phenomena could be followed by such viscosity studies of β-lactoglobulin at the oil/water interface. An increase in viscosity could be demonstrated, corresponding to disulphide bond formation in the monolayer (Dickenson, 1992).

The emulsification process is very rapid in relation to the time needed for a protein to reach conformational equilibrium at the oil-water interface. Measurement of the adsorption by ellipsometry or of the interfacial tension according to the drop/volume technique shows that it usually takes about 20 minutes until a plateau value has been reached (*cf.* Wahlgren and Arnebrant, 1991). This means that the protein molecules have various conformations at the oil/water interface after the emulsification process, and various relaxation phenomena will occur afterwards.

The degree of reversibility of protein adsorption is quite different when the two types of proteins shown in Figure 8.3 are compared. In some cases of globular proteins, complete reversibility has been demonstrated (MacRitchie, 1985). The main reason behind irreversibility, and this is the normal situation, is the occurrence of many contact positions between the peptide chain and the oil surface. Dynamically there will be times when a particular contact region is detached. However, it is very unlikely that all contact points will be detached at the same time, and this is the requirement for desorption. It should be pointed out that this is the main difference between polar lipids and proteins at the oil/water interface; lipids adsorb reversibly. The behaviour of different types of proteins at the oil/water interface, and their emulsification properties, have been reviewed recently (Tornberg *et al.*, 1989).

3. Lipid-Stabilized Emulsions

We have already described how an interfacial film of polar lipids can stabilize an emulsion (Figure 8.1). The presence of charged lipid species improves the stabilization of o/w-emulsions. In most cases when polar lipids are used in the production of emulsions, they form a separate phase at the interface. A structural requirement is that these phases are lamellar, with the possibility of exposing a hydrophobic surface towards water. There are three organizational alternatives of polar lipids providing this type of emulsification, as will be described below.

(*i*). *The* L_α*-Phase Interface Zone.* Non-ionic or zwitter ionic lipids give values for the water-layer thickness of about 20 Å, and such an emulsion interfacial layer was shown in Figure 8.1. Depending upon the o/w interfacial area, and the amount of L_α-phase available, there will be a varying number of lipid bilayers at the interface. The mechanical properties, and therefore the emulsion stability, is improved with increasing number of L_α-bilayers. The stabilization mechanism is the same as that behind formation of the L_α-phase, with the hydration force (see Chapter 3) providing repulsion between the oil droplets.

It should be stressed that as soon as an oil containing PC is exposed to water, an L_α-phase starts to build up at the interface. There are numerous reports in the literature of the effect of phospholipids on the oil/water interfacial tension. These measurements have hardly any meaning, as the phase formed at the interface has been neglected (*cf.* Söderberg, 1990). The interfacial tensions at the oil/L_α- and at the L_α-/water interfaces must both be considered.

When charged lipid species are present, the water layer thickness in the corresponding L_α-phase increases, and values of several thousands Ångstroms are common. Under these conditions, it is most likely that there will only be a monolayer of the polar lipid at the interface.

(*ii*). *The Gel-Phase Interface Zone.* A considerable stabilization effect of an emulsion stabilized by an L_α-phase interfacial layer can be achieved by crystallization of the L_α-phase into a gel-phase (Larsson, 1978). The

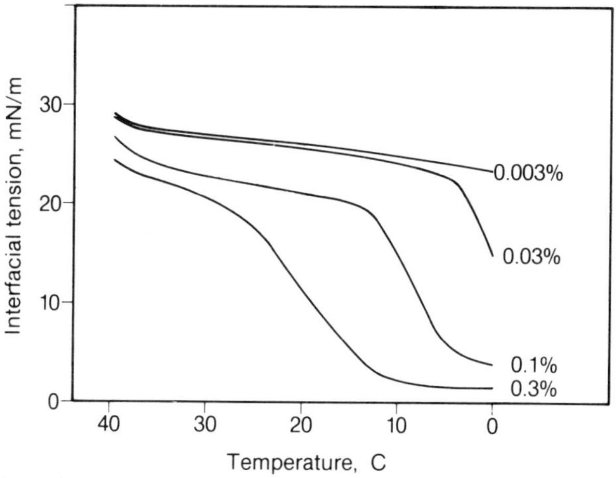

Fig. 8.4. Effect of temperature on the interfacial tension between water and oil of monostearin, after Krog and Larsson (1992).

emulsification process should then be run continuously while the emulsion is cooled through the transition $L_\alpha \rightarrow$ gel. A disadvantage with such gel-types of interfacial layers is that they are mechanically fragile, however.

(*iii*). *An Interface Formed by Surface-Active Lipid Crystals.* The dispersion of hydrophilic lipid crystals in water described in Chapter 7 can in fact emulsify an oil. The reason is the weak forces between lipid bilayers over the methyl end group gap. By shear only, these crystals can be opened, and if an oil phase is present at the openings, the new hydrophobic surfaces created can encapsulate the oil. It is therefore natural to term such crystals surface-active (Larsson, 1992c).

Industrial emulsification procedures for foods and for pharmaceutical products usually involve stirring under cooling. An emulsifier type that is often used is saturated monoglycerides. The following evidence shows that the monoglyceride forms surface-active crystals by such a process at the interface (Krog and Larsson, 1992).

The interfacial tension of a sunflower oil containing varying concentrations of monostearin is shown in Figure 8.4. It can be seen how the interfacial tension on cooling is drastically reduced at a certain critical temperature (T_γ). The variation of this temperature with monoglyceride concentration in the oil is shown in Figure 8.5. The shape of these curves is similar to the phase separation lines from the liquid state in the corresponding binary system oil-monoglyceride. A comparison with such a phase diagram, as illustrated in Figure 8.6, shows this similarity and confirms that the proposed explanation of the critical reduction in interfacial tension is formation of surface-active crystals at the interface. These crystals were also identified by X-ray diffraction.

(iv). *Lipid-Protein Competition at the Interface - A Mechanism Utilized in*

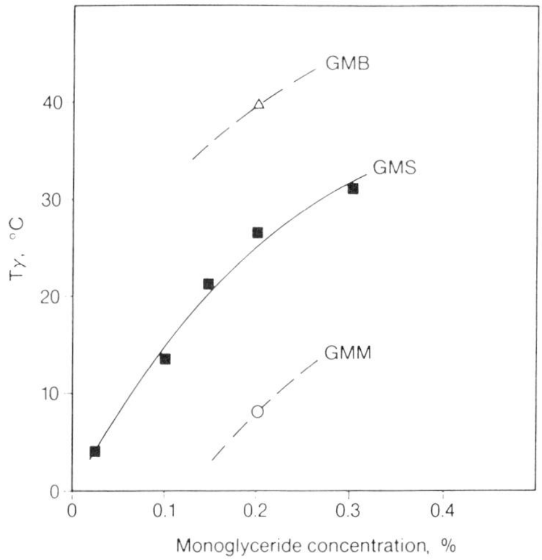

Fig. 8.5. Critical temperatures of reduction of interfacial tension versus concentration of monomyristin (GMM), monostearin (GMS) and monobehenin (GMB), after Krog and Larsson (1992).

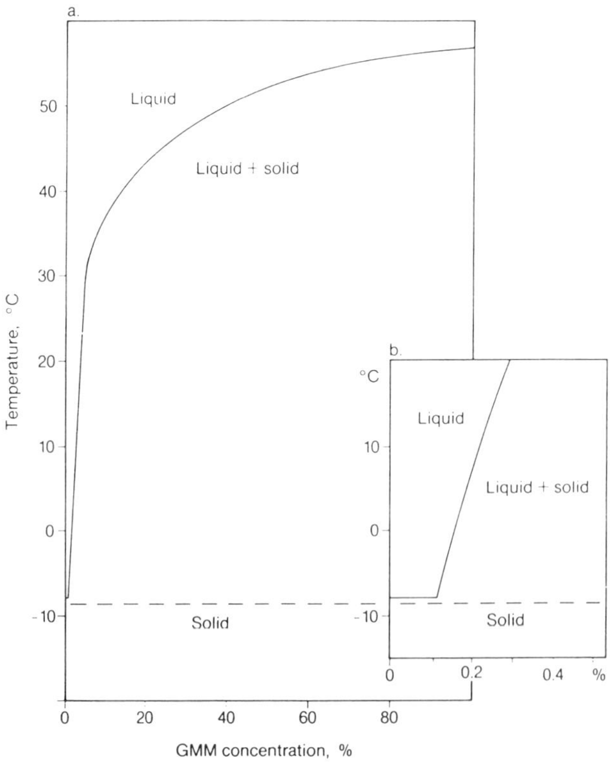

Fig. 8.6. Binary system of monomyristin-sunflower oil, after Krog and Larsson (1992).

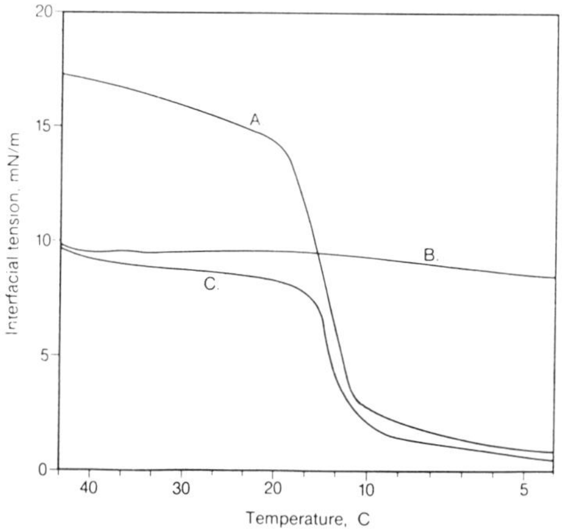

Fig. 8.7. Interfacial tension versus temperature of A, 0.2% (w/w) of glycerol monostearate in sunflower oil towards distilled water, B, the same oil against 0.01% (w/w) milk proteins in water and C, 0.2% glycerol monostearate in oil against 0.01% milk proteins in water, after Krog and Larsson (1992).

Ice Cream. When there is a competitive adsorption of a polar lipid and a protein at an oil/water interface, the polar lipid will usually win, as it reduces the surface energy more than does the protein. However, kinetic factors can change this situation in connection with emulsification. As proteins often adsorb in an irreversible way, films formed first by proteins at the oil/water interface may remain there, even if lipids are present that can lower the surface energy further. If the polar lipids are supplied in the water phase, where their monomer concentration is usually very low, water-soluble proteins will first fill up the interface. Whether or not the lipid molecules can then squeeze out the protein molecules from the interface is related to the degree of irreversibility in the adsorption and to stabilization of the protein monolayer by cross-linking.

It must be mentioned that polar lipids are sometimes used to destabilize a protein-based emulsion. One example is ice cream (further discussed in Chapter 14), where the oil droplets should undergo coalescence when the ice crystals melt. A model system corresponding to an ice cream emulsion is shown in Figure 8.7.

With a milk protein solution, the interfacial tension is 9.5 mN/m at 40°C with minor changes during cooling. The oil containing 0.2% monostearin shows the critical temperature behaviour of the interfacial tension towards water at about 18°C. The change corresponds to a surface-active crystal layer at the interface as discussed above. If this polar lipid is present in the oil in contact with the milk protein solution, the same temperature effect is seen on cooling. The final surface tension shows that the crystalline monolayer has squeezed out the proteins from the interface. The desired destabilization of the ice cream emulsion is obtained in this way (Krog and Larsson, 1992). It can be noticed that an emulsion with lower interfacial tension can be less stable as illustrated here. The strong steric stabilization of an emulsion by caseins is lost when the monoglycerides take over the interface.

C. WATER-IN-OIL EMULSIONS

This type of emulsion is rarely described in the literature, and there are few studies of systems which are w/o-emulsions in the strict meaning of the term. Butter and margarine are sometimes given as examples of w/o-emulsions, but this is not entirely correct. They are stabilized by the presence of crystals in the continuous phase, as mentioned earlier. Cosmetic formulations with dispersed water droplets as well as food w/o-emulsions are often of this type also, *e.g.* suspensions and emulsions at the same time.

Aluminium and calcium soaps have been reported to be effective stabilizers of w/o-emulsions (Busch and Neuwald, 1973). As such emulsifiers are insoluble in the oil as well as in the water phase, it seems obvious that the oil/water interface consists of surface-active crystals (or gel-phase). From a curvature point of view, there is no reason for preference for o/w- or w/o-emulsions when the interface consists of surface-active crystals. Whether an o/w- or a w/o-emulsion is formed depends upon the relative proportions of the two phases and their viscosity. From a stability point of view, however, the two emulsion types are quite different. Electrostatic repulsion cannot be used to stabilize w/o-emulsions, and there are few known polymers that can provide steric stabilization in the oil phase. The presence of crystals in the oil phase provides one of a few methods for stabilizing these emulsions.

Wool wax ("lanolin") is a common component used to stabilize water-in-oil emulsions for skin care formulations. The wool wax is the lipid fraction extracted from raw wool. It consists mainly of fatty acid esters of sterols and fatty alcohols. An even better emulsifier for w/o-emulsions is obtained if the wool wax alcohols are separated. They consist of two sterols mainly, of which cholesterol is most abundant, and about a third of long-chain *n*-alcohols (Hoppe and Larsson, 1981). A wool wax alcohol w/o-emulsion is obtained by emulsifying at a temperature where the wool wax is liquid, and then by cooling the mixture under continuing emulsification. The phase transitions during cooling have been analysed by X-ray diffraction (Hoppe and Larsson, 1981). In the polarizing microscope, it can be seen that a birefringent phase is formed at the oil/water interface on cooling. Without any water present, the same shape and birefringence as obtained from liposomes were observed. This phase is crystalline and consists of bilayers of cholesterol (and other sterols) according to the X-ray diffraction pattern. The emulsion is thus stabilized by a film of surface-active crystals (mainly cholesterol) at the interface. At lower temperatures, the aliphatic alcohols crystallize in the bulk of the oil phase.

When the wool wax alcohol fraction is used in stabilized w/o-emulsions, which can also contain other oil components (triglyceride or paraffin oil), it can be concluded that sterol multilayers at oil/water interfaces provide the stabilizing effect. If the content of liquid oil is increased, it is necessary to immobilize the oil phase by crystals. They should be hydrophobic, not to interfere with the interface, and a solid wax is often used.

Crude oil is a mixture of aliphatic and aromatic hydrocarbons with some polar components, carboxylic acids, phenols and asphaltenes. The polar components are responsible for w/o-emulsion formation in crude oil reservoirs, and such emulsions have a high viscosity which complicates pipe-line transportation. Crude oil emulsions have therefore been studied in detail (*cf.* McMahon, 1992). The size of the wax crystals has a major influence on the stability of these emulsions, and it has been shown that they do not act at the interface but in the bulk of the oil phase (McMahon, 1992).

There is a report of a special emulsifier favouring w/o-emulsions because of the curvature at the oil/water interface (Chattopadjyay *et al.*, 1992).

D. DEMULSIFICATION

As mentioned at the beginning of this chapter, flocculation and coalescence are the two processes which limit the lifetime of an emulsion. These phenomena will now be considered in detail in order to provide some guidelines for emulsion breakdown, demulsification. In certain situations, emulsions are not desired, and efficient technology is needed to separate oil from water. One example is waste-water from the dairy and slaughter industries. Another is lipid extraction of water-containing materials.

1. Flocculation and Coalescence

Flocculation is the reversible aggregation phenomenon due to the secondary minimum in interaction energy between the particles, as shown in Figure 8.1. The mechanism leading to coalescence involves thinning of the liquid film between the particles and finally film rupture. The various factors influencing film drainage have been summarized in an excellent review (Menou and Waran, 1985). The interfacial viscosity is an important factor at levels between 10^{-6} and 10^{-3} Pa m sec, and outside this range it has no effect. In this viscosity range, the drainage rate is decreased by:

- high values of the gradient of interfacial tension with respect to concentration

- high drop viscosity
- low drop phase diffusivity
- low interfacial diffusivity
- surfactant solubility in the continuous phase

The probability of film rupture increases with film thickness once the film has thinned to a critical level.

2. *Demulsification*

We learned from the DLVO-theory that the electrostatic repulsion is reduced by the addition of electrolytes. An increase in temperature acts in the same direction, by increasing Brownian motion and decreasing the viscosity of the continuous phase, and therefore of the film between flocculated particles. *Chemical* demulsification can be based on these factors or on the addition of surface-active molecules, that can displace the emulsifier molecules from the interface. These molecules must not be able to form stable interfacial films. *Physical* separation methods are frequently used for demulsification. The most important techniques are centrifugation, filtration and air flotation. Centrifugation and filtration act by forcing the droplets close together. In air flotation, the emulsion droplets are enriched in the foam and are therefore forced together (Menou and Wasan, 1985).

E. FAT CRYSTALLIZATION IN EMULSIONS

In the discussion on crystal nucleus formation in Chapter 2, it was stated that fat crystal nuclei are usually formed by the presence of active sites (heterogeneous nucleation). When an oil is emulsified, these active sites will be diluted, and most droplets will be free from such active sites. Crystallization can then only start by homogeneous nucleation. In studies of tristearin crystallization in oil, it was observed that the crystallization temperature dropped by about 20°C on emulsification (van den Tempel, 1968).

The functionality of fat droplets is closely related to the surface area of fat crystals, and this can vary enormously with crystallization temperature. The higher the supercooling, the more nuclei are formed, and it has been shown that the average size of fat crystals can differ by a factor of 10^5 by varying the nucleation temperature (van den Tempel, 1968).

An additional complication in emulsion crystallization is differences in the polymorphic behaviour compared to that in bulk. The polar lipid monolayer at the surface of the oil droplet may promote a particular crystal form. A study of crystallization of milk fat in bulk and in the fat globules of milk showed significant differences (Söderberg *et al.*, 1989). The triglycerides in the milk fat globule crystallize directly into a β'_2-form, whereas they form a β'_1-form in a continuous fat phase.

F. EMULSION STRUCTURE AND ENZYMATIC FAT DIGESTION

The structure of emulsion droplets in some foods and in emulsions for parenteral nutrition (usually prepared from soybean oil and phospholipids from egg yolk or soybean oil) have an interface just like that shown in Figure 8.1. How is it then possible for the lipase to reach the triglyceride molecules? Intestinal lipolysis, for example, is a remarkably fast process; gram-quantities of lipid per cm are absorbed during passage through the upper intestine.

Hamilton and Small (Hamilton and Small, 1981; Hamilton, 1989) have determined the limit of solubilization of triglycerides in the L_α-phase of PC. Triolein has a solubility of about 3%, whereas the solubility is higher in the case of medium-chain triglycerides. The lipase, which can only act when it is adsorbed at the interface, will first form 2-monoglycerides and free fatty acids of triglyceride molecules in the surface zone bilayers. There is an interesting aspect of the monoglycerides produced in this context. They are known to tend to fuse polar lipid bilayers or membranes, a feature related to formation of a cubic phase. If this takes place on the emulsion surface, there will be a "hole" where lipase can reach the triglyceride core. When a triglyceride oil droplet is digested, the occurrence of a surface zone of the cubic phase (Patton and Carey, 1979) may even indicate that the cubic structure plays an active role in lipolysis (Larsson, 1989). It should also be noticed that the triglyceride molecule, when entering the active site of the enzyme, must have all chains in the same direction; thus they possess a conformation different from that in bulk. This demonstrates the significance of the interface.

Hamilton (1989) also determined the solubilization of tripalmitin compared to triolein in PC bilayers. The remarkable result was that tripalmitin was soluble to the same extent as triolein in the bilayers (in the temperature range 30-50°C). In this liquid-crystalline bilayer, both triglycerides therefore have the same disordered conformation, although they are liquid or solid respectively in bulk. This result may explain the fact that fats solid at body temperature are also absorbed effectively in the intestine.

The difference in solubility in PC bilayers between long-chain and medium-chain triglycerides was mentioned earlier. In a comparative study of the hydrolysis of the corresponding emulsions, a solubility of trioctanoin of 11% was observed by Deckelbaum *et al.* (1990). They found a faster hydrolysis of the medium-chain triglycerides, although the enzyme affinity was lower, and they related this to the higher solubility in the PC bilayers.

A few words must also be said about protein-based emulsions, which occur frequently in foods. The ESP-values of triglyceride monolayers were discussed in Chapter 4, where it was shown that values up to about 20 mN/m can occur. Protein monolayers on water exhibit similar plateau-values at adsorption, although ESP-values seldom can be measured because of their solubility. This behaviour indicates that because of similar "strengths" in competition on an emulsion droplet surface, lipase will "see" many triglyceride molecules when entering such a surface.

Cholesterol esters are solubilized in a similar way in PC bilayers with the carboxyl group in contact with water. Therefore they can be attacked enzymatically like triglycerides (Hamilton *et al.*, 1991). Finally it should be pointed out that pancreatic phospholipase and lipase can both digest the phospholipid coat of emulsions.

CHAPTER 9

LIPIDS IN BIOMEMBRANES - DRUG PENETRATION THROUGH MEMBRANES - MEMBRANE MIMICKING

Structure and function of biological membranes represent one of the most active fields of research within the biosciences at present. The ambition behind this chapter is to deal with aspects of applications related to the lipid structure in membranes. Knowledge of the organization of the lipid bilayer is relevant in order to understand membrane penetration and the design of drug delivery formulations. There are also interesting recent results indicating that knowledge of the biomembrane structure can be used for mimicking this surface on artificial materials in order to design biomaterials.

To be able to discuss these aspects it is necessary to start by summarising the general structural principles of biomembranes. Particular attention will be given to cubic membrane systems, a new field which is changing our view of membrane functions and will probably result in significant new applications.

A. GENERAL STRUCTURAL FEATURES OF MEMBRANES

The self-assembly of polar lipid molecules into infinite bilayers with water on each side is the mechanism behind formation of cell membranes, and it is a feature fundamental to the development of organized life. The general model of biomembranes, the *fluid mosaic* model, introduced by Singer and Nicolson (1970), describes how proteins are solubilized in two ways, as "integral" spanning the bilayer or "peripheral" if they are loosely attached to either side of the lipid bilayer. The proteins have translational freedom along the bilayer similar to that in a two-dimensional liquid state. In fact, this model together with its functional aspects are direct consequences of the liquid-crystalline character of the lipid bilayer.

Membrane proteins have to contain sequences of amino acid residues which are hydrophobic in the region where they are embedded or penetrate the hydrocarbon chain core of the lipid bilayer. Many such proteins have now been characterized. A third alternative of protein attachment to the membrane has also been revealed. It takes place through phosphatidylinositol, which is covalently linked to an outside protein *via* a spacer arm.

It is obvious that the hydrocarbon chain conformation and mobility - the bilayer "fluidity" - can influence the conformation of membrane proteins. Therefore there must be an efficient regulation of membrane composition in order to control the fluidity. This has been confirmed by numerous studies of effects on lipid composition due to variations in environmental factors, *e.g.* bacterial membrane changes induced by temperature. A perfect temperature adaptation in the fluidity was observed in a recent study of the evolutionary adaptation of the bilayer in fish brain membranes from a wide range of species living in different thermal habitats (Behan-Martin *et al.*, 1993).

B. LIPID PHASE PROPERTIES AND MEMBRANE FUNCTIONS

Our understanding of membrane functions has reached a very high level during the latest two decades, because of our knowledge of protein structure in particular. However, the functional role of the lipids is still poorly understood. A great deal is known about chemical lipid changes, such as phospholipase activity leading to the formation of second messengers or formation of prostaglandins. The significance of the physical state of the lipid bilayer on the other hand remains to be determined. We saw in Chapter 3 how lipid molecular assemblies can undergo reorganizational changes of high complexity. Whether or not these are utilized in membrane functions is not well understood.

There are a few cases where a lipid phase transition has been demonstrated to be responsible for a particular function. Epithelial cells are joined by so-called tight junctions. From recent work, it has been concluded that the permeability of tight junctions can be described as lipid phase equilibria between the L_α-phase (open state) and the H_{II}-phase (closed state) (Hein *et al.*, 1992).

Another analogue to a lipid phase in three dimensions, that seems likely to be involved in membrane functions, is the cubic arrangement of the lipid bilayer. Two aspects of the cubic bilayer conformation in membranes are considered below. The first concerns membrane assemblies which have the same structure as cubosomes (see Chapter 7). The second aspect is the possibility of a phase transition involving an intrinsically curved bilayer as a functional state in membranes.

C. CUBIC BIOMEMBRANE ASSEMBLIES

The first proposal for the occurrence of a cubic IPMS-structure in biomembranes was an interpretation of electron micrographs of thylakoid membranes in their passive "storage" state, the prolamellar body (Larsson

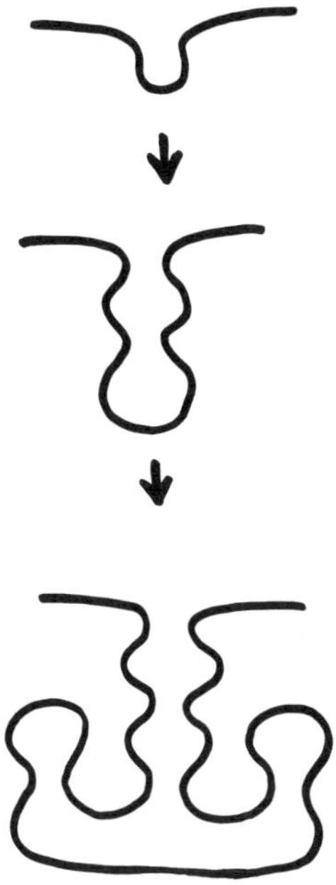

Fig. 9.1. Formation of a cubosome region of a membrane by invagination and successive transfer of bilayer material from the region of planar conformation to the cubic membrane assembly.

et al., 1980). This structure has now been analysed rigorously and proved to be a C_D-phase (Landh, 1994). The transition from the planar conformation of the membrane lipid bilayer into a cubic membrane assembly can take place as illustrated in Figure 9.1. If the bilayer curvature locally tends to be curved as a cubic bilayer, an invagination can be formed, and bilayer material can then be transported into the cubosome region.

These changes can also go in the opposite direction. An invagination forming a cubic bilayer structure can be used as a depot to increase the area of the membrane front. The local change in curvature behind such transitions might be a conformational change in a membrane protein or a contraction/expansion of a lipid polar head .

Landh (1994) has reported more than a thousand clear cases of cubic cell membranes under physiological conditions. A description focused on the role of curvature has recently been reported (Andersson *et al.*, 1994). We will consider three examples here.

When the t-tubular system of membranes through the muscle cells is depolarized by the action potential, calcium is released from the closely located sarcoplasmic reticulum. This leads to the actin-myosin contraction movement. It was shown by Landh (1994) that the t-system is a cubic membrane. This is obviously a most effective structure for penetration of the excitation membrane through the muscle cell.

Many endothelial cells exhibit a system of branching invaginations in the inter-endothelial cleft, appearing like vesicles in electron microscopy sections (Bundgaard, 1984). The morphology of these invaginations corresponds to a constant-curvature surface, and with our present knowledge of lipid bilayer phases, they can be described as a part of a cubic membrane.

A beautiful example of an "undulated" arrangement of a cell membrane was reported by Öhman (1974), and is shown in Plate 9.1. This structure was observed in epithelial cells of the retinal pigment, and it appears to be identical to an expected fracture plane of the cubic C_G structure (Landh, 1994).

There are simple analytical expressions which describe the cubic structures with good accuracy (but not exactly) (*cf.* von Schnering and Nesper, 1991). A good approximation to describe the C_P-phase is:

$$\cos x + \cos y + \cos z = 0$$

If there is a pressure gradient over the surface, the minimal surface is transformed into a constant curvature surface. With the average curvature equal to K, the surface is approximated by:

$$\cos x + \cos y + \cos z = K$$

When K is approaching 1, the channels become closed at the necks between the structural units, as can be seen in Figure 3.7. This means that a continuous change in the average curvature, from the minimal surface structure (with which we are familiar from the cubic phases) will end by the formation of free vesicles. As soon as the necks become closed, the structural units, which earlier bud out in the x-, y- and z-directions, will become spherical. The two formulae above show how changes only in the average curvature can result in transitions between a cubic phase and an arrangement of vesicles.

It is natural to expect that the mechanism described above explains how mass-cooperative fusions of vesicles can be controlled, such as the release of transmitter molecules at the presynaptic membrane (Larsson and Andersson, 1986).

D. A BIOMEMBRANE PHASE TRANSITION INVOLVING L_α AND CUBIC BILAYER CONFORMATIONS

There are numerous observations on periodic curvature in single membranes from animal tissues (*cf.* Hirsch *et al.*, 1988) and from Gram-positive bacteria (*cf.* Sternberg *et al.*, 1987). In cases of double membranes, there are similar reports (*cf.* Lane and Harrison, 1979)

A two-dimensional analogue of a cubic lipid-water phase can be formed by sectioning out a single layer by two parallel planes. There are two fundamental two-dimensional minimal surfaces formed in this way. One is parallel to the (111)-plane of the D-surface and the other is parallel to the

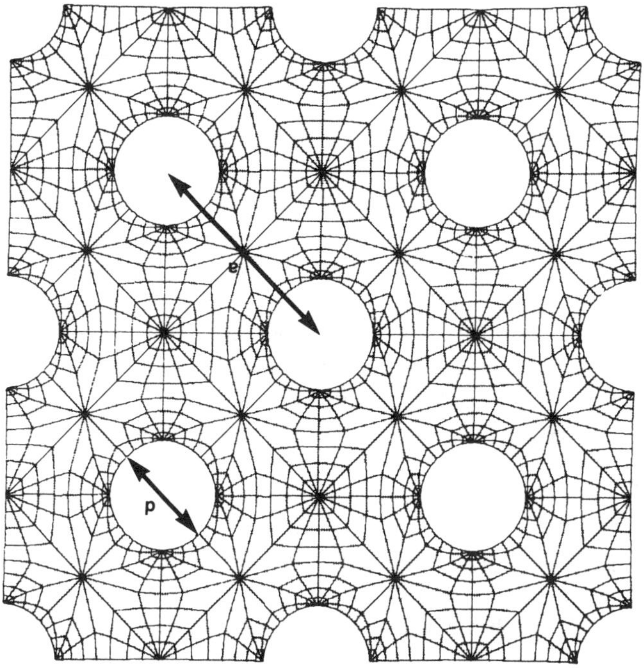

Fig. 9.2. A perpendicular view of a hypothetical membrane formed by a (111)-section through the C_P-phase. Two-fold axes and levels of constant Gaussian curvature are indicated, after Larsson (1988).

(100)-plane in the P-surface. The sectioning planes will then form "holes" where they pass through the surface. These holes are assumed to contain protein units in corresponding membranes. It was proposed that this membrane model is involved in membrane phase transitions, as will be discussed below (Larsson and Andersson, 1986; Larsson, 1988). One alternative of such a membrane with an intrinsic periodic curvature is shown in Figure 9.2.

The differential geometry of such a two-dimensional minimal surface has been analysed, and it was shown that a transformation corresponding to continuous variation of the curvature towards the planar bilayer can take place as an extreme (Hyde *et al.*, 1986). The proteins of the curved bilayer have to be organized, and there is experimental evidence for the occurrence of such membrane structures (Larsson, 1989).

A periodic curved bilayer model of a membrane would be able to accommodate varying amounts of lipid bilayer per unit front of membrane, a need related to membrane transport involving vesicles. A change between planar and curved bilayers has interesting implications. The propagation of such a phase transition "wave" can move with the speed of sound and requires an extremely low energy input. Such transitions can provide a simple possibility for communication along the surface of a membrane.

An interesting functional aspect of this hypothetical biomembrane phase transition is the possibility of direct control of the bilayer composition by on/off-switching of the membrane-bound enzymes responsible for lipid synthesis or modification (Larsson, 1988). The lipid composition in most membranes appears to keep the bilayer balanced on the edge of a phase transition, from the L_α-conformation towards one corresponding to the cubic or H_{II}-structure. Thus variations in temperature, pressure and so on are directly reflected in the composition of membrane lipids (so that they keep close to this transition). Consider for example the methyltransferases responsible for converting PE to PC. The PEs of unsaturated lipids can form cubic phases (Seddon, 1990), whereas PCs favour the L_α-conformation. The $L_\alpha \rightarrow C^{2D}$ phase transition could provide the switch-mechanism for control of this enzyme reaction.

A lipid bilayer with the L_α-conformation corresponding to the biomembrane structure can exhibit a phase separation with segregation of different lipids into each of the two fluid regions. Such a phase separation has been demonstrated in black lipid film model membranes consisting of phosphatidylglycerol mixtures exposed to calcium ions (Knoll *et al.*, 1991). A lateral organization of the bilayer in the L_α- and cubic conformation is one alternative for such lateral phase separation in a biomembrane.

E. GLYCOLIPIDS IN RECOGNITION PHENOMENA.

Sugar groups exposed on biomembranes have functions in cellular recognition phenomena. The various possibilities of this system code are enormous. As an example, we can consider a group consisting of three different monosaccharides; they can be combined in more than one thousand ways because of variations in binding position and configuration. If we compare a peptide consisting of three different amino acids, they can only be combined in three ways.

The sugar groups exposed on the surface are linked to membrane lipids and membrane proteins; glycolipids and glycoproteins. Glycolipids are located in the outer monolayer of the plasma membrane, and they are often involved in adhesion phenomena such as microbial adhesion. The sugar groups in animal membranes are linked to sphingolipids, whereas they are attached to glycerolipids in plant membranes. For example, a certain *E. coli* strain can adhere to glycolipids at the intestinal wall if the surface exposes a disaccharide of galactose and sialic acid. This mechanism explains the remarkable specificity of infections. There are many recognition sites (adhesins) on the bacterial surface, and adhesion may require more than one receptor and even different receptor structures.

An interesting aspect of these very polar membrane lipids is their dynamic properties. It is sometimes forgotten that the lipid molecules in the membrane are mobile not only *within* the membrane. A surface membrane will always lose its lipid molecules until the lumen concentration is saturated. In the case of phospholipids, with solubilities less than about 10^{-10} M, this effect is negligible. Glycolipids on the other hand, which can contain 20 to 30 sugar groups, must have a considerable solubility. Therefore the loss into an outside water layer will require a corresponding renewal of the membrane lipid bilayer. This aspect of the gastrointestinal membranes has been neglected and will be discussed in Chapter 12. Such lipid molecules should also be expected to move in and out in relation to the bilayer plane, so that on the average a few carbon atoms of the hydrocarbon chain are located outside the hydrocarbon chain core of the bilayer. These dynamics might be a significant factor in the adhesion process, helping the sugar groups to find their receptor sites.

F. DRUG PENETRATION THROUGH MEMBRANES

In order to reach its site of action, a drug must penetrate various barriers, the most common one being cell membranes. Small molecules will leak through the membranes, and penetration values for different types of ions and small groups through the lipid bilayer are given in Chapter 13. Larger molecules have to be amphiphilic in order to penetrate. A general surface property of membranes under physiological conditions is a negative net charge. A positive net charge in a drug molecule will therefore increase the tendency to reach the membrane. However, penetration through the hydrophobic core takes place more easily in the case of non-ionic molecules. Penetration through epithelial cells can occur both through their plasma membranes and through the tight junctions, as discussed in Section C above.

The *blood-brain barrier* represents an extremely efficient membrane function, which also exhibits an enormous selectivity towards the penetration of drugs. It restricts the passage of substances from the blood to the brain, and through this barrier the brain is protected from variations in the composition of the blood plasma. The physical site of the blood-brain barrier is the endothelial cells. These cells are laterally joined by tight junctions, and the membranes provide the barrier. There are clear observations of membrane invaginations that either can be open for penetration towards the luminar side of the capillaries (towards the blood) or towards the brain (Bundgaard *et al.*, 1979). These features are fully consistent with a cubic arrangement of the membrane as discussed above. A thin zone of the cubosomal structure, *cf.* Chapter 7, can function as such a lock; by a simple curvature-driven transition, "holes" towards the blood side can be closed at the same time as holes towards the lumen side are opened. These endothelial cells have very high electrical resistance, and this means that the ion permeability is low. Another unique feature of the capillaries of the blood-brain barrier is their high metabolic activity, as evident from a high content of mitochondria, considerably higher than in capillaries from other tissues (Johansson, 1990).

Hills (1989) has reported an electron microscopy study indicating that the physical identity of the blood-brain barrier comprises a number of lipid bilayers, and this structure seems consistent with a cubosomal layer. Furthermore, based on contact angle measurements, it was reported that there is a hydrophobic surface. This conclusion will be discussed in connection with the claimed hydrophobicity of the gastric wall (Chapter 12, Section B).

The degree of penetration of a drug through the blood-brain barrier shows a close correlation with its lipid solubility. Exceptions are glucose, the fuel of the brain, and certain amino acids with special carrier/transport systems. With increasing molecular weight, the penetration of the drug is reduced.

It has been observed that cationization and glycosylation can result in permeability of proteins (Johansson, 1990). Cationization probably increases permeability by a mechanism related to penetration of polycations, *i.e.* by reducing the electrostatic repulsion of the membrane surface.

Numerous strategies to pass peptide drugs through the blood-brain barrier have been reported, and a promising "molecular packaging" technique was recently described (*cf.* Borman, 1992). Cholesterol is linked to the peptide to increase the lipid solubility, and a targeting group is also attached. This molecule is neutral before passage, and it becomes positively charged on the brain side due to an enzymatic reaction in the brain. As positively charged molecules cannot pass out through the barrier, the peptide is trapped.

There are more drastic ways of increasing this membrane permeability, such as mechanical stress or osmotic opening of membranes (Johansson, 1990).

G. BIOMATERIALS MIMICKING MEMBRANES

Within the extensive research on medical implants and extracorporal treatment of blood (such as dialysis), we have now a rather good understanding concerning the requirements for surface structures of materials exposed to a biological tissue or fluid. When any foreign material comes into contact with blood or a tissue, conformational changes of proteins at the interface will always take place, and the extent of these changes will determine whether this material is acceptable (biocompatible). Biocompatibility can be achieved by surface modifications, and heparinization of surfaces to reduce blood coagulation is a well-known example. A new approach, which seems promising, is to add a surface layer so that the material will appear like a cell membrane from the outside.

It seems obvious that a biomaterial with a surface like a cell membrane should be as close as we can ever come towards biocompatibility. It is also quite obvious that there are many complications on the way to developing such a membrane-mimicking surface. Chapman and coworkers have been working with this approach for longer than a decade (*cf.* Chapman and Charles, 1992). Their choice of lipid was phosphatidylcholine; the predominant lipid in human cell membranes. There is constantly a wide spectrum of chemical changes of membrane lipids, in particular by the activities of phospholipases A_2, C and D. This is probably the main obstacle to be overcome in order to obtain such surfaces which are inert over long periods.

Direct deposition of phospholipids on solids by the Langmuir-Blodgett technique (see Chapter 5, Section A) has been tried, but such a surface layer is too fragile from a mechanical point of view. The first more promising solution was achieved by polymerization of deposited mono- or multilayers of the phospholipid. One route of cross-linking was based on triple bonds in the acyl chains, which can form lateral bonds on exposure to UV-light (Hayward and Chapman, 1984). The disadvantage is that it is hardly possible to obtain a network involving every molecule. Small aggregates, dimers, trimers and so forth, can therefore leave the surface and may cause toxic reactions.

The best solution up to now has been direct covalent attachment of the polar head group to a polymeric surface allowing such chemical linkages. Remarkably good performances have been reported for such surface-modified materials, particularly with regard to blood contact (Hayward and Chapman, 1984).

A similar approach has been reported by Ishihara and coworkers (1990). They obtained reduced thrombogenicity by a surface coating of poly(2-methacryloyloxyethylphosphorylcholine). After surface polymerization, a zwitter ionic exterior, similar to the PC polar head sheets, is obtained.

As a final comment it can be mentioned that the mobility of individual lipid molecules in biomembranes cannot be mimicked in the surface modifications described above. This mobility is certainly a significant factor in bioadhesion processes also.

CHAPTER 10

LIPIDS IN LUBRICATION

Oils are used in numerous metal working processes in order to cool and lubricate. Mineral oils are often preferred as they have better chemical and microbial stability. The use of vegetable oil is increasing, however, because of environmental considerations.

As well as the liquid state, lipids in the solid state and in the form of liquid-crystalline phases are used in lubrication.

A. LUBRICATION

Lubricants are able to reduce the friction between solid surfaces. Friction (F) is the force tending to oppose movement. Between two solids moving relative to one another it is equal to:

$$F = \mu W$$

where μ is the coefficient of friction and W is the load (the force at the contact region normal to the direction of movement). In order to start the movement, it is necessary to overcome the adhesion, which is equal to the product of the contact area and the shear strength of the junction. A deformation can then occur in the contact zone, and plastic and/or elastic deformation gives a contact area that is related almost linearly to the total load.

When there is a liquid film between the solid surfaces and no deformation occurs, *hydrodynamic lubrication* is obtained. The friction, when there are no points of contact, is related only to the viscosity of the fluid. The film thickness is proportional to the viscosity (η) and the movement velocity (v) according to:

$$(\eta v)^{2/3}$$

From the viscosity, it is therefore possible to estimate when contact will occur. Even when there is contact between the two solids, the friction can still be much lower than what might be expected from the coefficient of friction between the solids. The lubricant monolayer on each solid takes up a part of the load. The force needed to move the solids then depends on the area of solid-solid contact in relation to the area where there is contact between the lubricant monolayers.

B. LIQUID-CRYSTALLINE PHASES

The lamellar and cubic phases will be discussed as both possess significant lubricating properties.

The L_α-phase often has dual functions in food applications, to provide lubrication and to stabilize dispersions. One example is chocolate, discussed in the food chapter. The best lubrication is obtained when the L_α-phase forms an oriented layer parallel to a solid surface.

The lubricating effect of the L_α-phase can be demonstrated by allowing a droplet of this lipid-water phase to orient with time between two cover-slips. The orientation can be followed in the polarizing microscope, as the birefringence is successively lost when more and more of the bilayers become parallel with the cover-slip. The L_α-phase has least resistance against flow along the methyl end group gap, where there are very weak van der Waals forces and liquid-like disorder. The L_α-phase can also exhibit flow along the water layers. In contrast, if the shear velocity is too high, the bilayers will successively adopt orientations other than that parallel to the flow, and the viscosity will increase. A disordered L_α-phase can even show some degree of elasticity at small deformations.

Before the cubic symmetry was known, cubic phases were sometimes termed "viscous isotropic" according to their rheological properties. Cubic lipid-water phases show viscous deformation, and elastic behaviour only at extremely small deformations. This is natural with regard to their structure. In fact, the cubic phase can be regarded as a crystal, as it is periodic in all three dimensions, contrary to "normal" liquid-crystals. The cubic phase shows a remarkable lubricating effect when excess of water is present, as can easily be demonstrated. The cubic phase monoolein-water phase, for example, can be spread as a film with good adhesion on a fingertip. By rubbing the thumb against this fingertip in running water, the lubrication can be felt, and it lasts a remarkably long time. The amount of water needed to remove the film indicates that the phase disappears from the lubricating region only because of the solubility of the monomer (about 10^{-6} M).

The lubricating effect must involve an aqueous surface film between two opposing cubic layers, as the cubic phase itself shows a strong resistance against flow. The cubic phase at the air/water interface is in equilibrium with a condensed monolayer phase. The equilibrium spreading pressure is about 40 mN/m, and the molecular conformation of the monolayer is like that in the L_α-phase. This indicates that an interfacial film with an L_α-conformation of a monoolein-water cubic phase is formed during lubrication. The interface between the cubic phase and an outside water phase should be expected to look just like the surface of a cubosome (see Chapter 7). Formation of an interfacial L_α-like bilayer structure along the plane of flow is therefore proposed as an explanation of the lubricating behaviour of the cubic phase. It seems likely that there will be some applications of this kind of lubrication mechanism, particularly with regard to its water-compatibility.

Fatty acid salts of divalent ions are sometimes used as aqueous lubricants. The rheological properties of such water dispersions have been analysed (Burg and Jeschke, 1992). From a structural point of view, such dispersed aggregates will be lamellar liquid-crystals or gel phases depending on chain length and unsaturation.

Phospholipids appear to play an important role in joint surface lubrication, and occur in the synovial fluid (Hills, 1989).

C. CRYSTALLINE PHASES

The crystal structure of simple lipids with all of the methyl end groups in one plane explains their effectiveness as solid lubricants (*cf.* Figure 2.1.). Because of the weak van der Waals forces over the methyl end group gap, the bilayers can slide easily along one another.

Fatty acid salts are frequently used for lubrication in powder processing, for example in the production of tablets by a pressure process involving the actual drug and carriers. Divalent salts, such as zinc, calcium or magnesium are used, as they give crystals with high melting points. From a functional point of view, it is important that the crystals have a large surface area parallel to the bilayer plane compared to the crystal thickness. Other applications of these salts involve the metal-working industry.

The art of greasing skis is an example of solid lubricant development, which has reached a high degree of sophistication. Waxes in the α-form are often involved, and a surface of rotationally disordered methyl end groups is obtained. A major requirement is a non-adhesive surface with regard to the contact zone between the ski and the snow, *i.e.* ice crystals and an interfacial water film that might be formed as a result of the load.

D. GREASE AND PLASTIC FATS

Grease is a gel structure quite analogous to that of gelatine. A continuous mineral oil phase is immobilized by a network of surfactant molecules associated into fibres. A detailed study has been reported of a grease formed by lithium 12-hydroxy-stearate in mineral oil, using electron microscopy (Prost and Rondelez, 1991) and X-ray diffraction (Terech *et al.*, 1985). This hydroxy fatty acid is prepared from hydrogenated castor oil. The gel network consists of entangled ribbons. The cross-section structure of the ribbon is shown in Figure 10.1. It consists of a lipid bilayer strip; the limited lateral extension is probably related to the divergence of the hydrocarbon chain. The ribbons are twisted into a helical conformation.

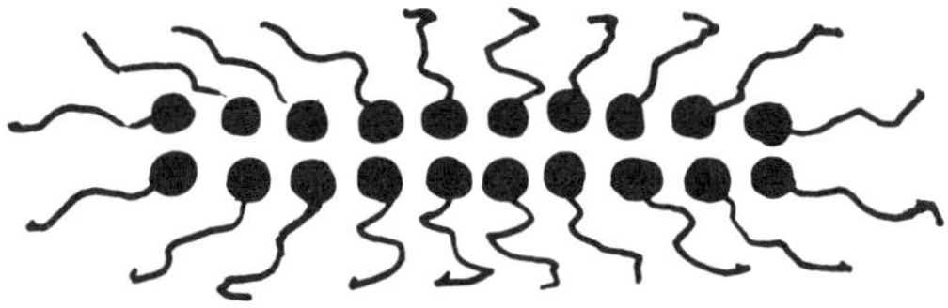

Fig. 10.1 Schematic illustration of the cross-section of the helical ribbon structure of lithium 12-hydroxy-stearate in mineral oil.

There is a now a trend to use plastic fats and vegetable oils immobilized by fatty acid salt aggregates instead of grease, because of environmental considerations. Examples of recent applications are lubricating formulations for agricultural or forestry machinery. It is obvious that spills of such materials are metabolized faster than those of conventional lubricants.

CHAPTER 11

THE ROLE OF LIPIDS IN HUMAN SKIN - SKIN CARE FORMULATIONS AND DRUG PENETRATION

The skin provides mechanical protection to our outer tissues, and it functions as a barrier against penetration of chemicals. A layer of dead cells, the stratum corneum, forms a surface layer with the viable epidermis underneath. The so-called papillary layer forms the contact zone between epidermis and dermis. The thickness of the stratum corneum is a few μm, whereas that of the epidermis is about 100 μm. Yet, the stratum corneum alone is responsible for the barrier function, an effect due to its content of lipids. The area of the skin of an adult is about 2 m^2, and the water evaporation through this surface (not taking the sweat-glands into account) is only about 300 g per day (*cf.* Forslind, 1994). Control of water evaporation is the most important function of the stratum corneum. Without this water barrier, we could not survive.

A. SKIN LIPIDS

Intercellular lipids play a fundamental role in the barrier function of the skin. Surface lipids originate from two sources, the sebaceous glands and cell membrane residues from the differentiation process leading to formation of epidermis. The whole skin is renewed within about one month. Sebaceous glands secrete a hydrophobic oil, sebum, which contains predominantly triglycerides (about 58%), wax esters (26%) and squalene (12%) (*cf.* Greene *et al.*, 1970). If lipids are extracted from regions rich in sebaceous glands, such as the forehead and scalp, more than 90% of the lipids are derived from sebum.

Covalently bound lipids have recently been identified in stratum corneum of porcine epidermis (Chang *et al.*, 1993). They consist of ω-hydroxy-ceramides, and they amount to 2.4% of the dry weight of the tissue.

We know from Chapter 5 that triglycerides exert a spreading pressure when exposed on a hydrophilic surface. This explains the spreading of sebum to give a protective surface film outside the skin surface, that is produced constantly. A simple demonstration of the interfacial properties of this film can be performed. Fill a vessel with a diethyl ether-saturated water solution, for example a trough like that used in the surface balance (Chapter 5). A small amount of sebum can be taken from the forehead by a clean needle. The diethyl ether/water solution can be lit by a match. If the needle is then put in contact with the water surface, it is possible to follow how rapidly the spreading of the sebum surface film will extinguish the fire. The mechanism behind this effect is the reduced evaporation of diethyl ether because of the presence of the monolayer.

The sebum triglycerides are more or less hydrolysed, and this results in a pH approaching 4. The antimicrobial activity of the free fatty acids is an important physiological function of the sebum film.

The other source of skin lipids is epidermal lamellar bodies and cell membrane residues from the keratinization process. These lipids form a lamellar structure in the intracellular space of stratum corneum. This surface layer of the skin can easily be scraped away or stripped off by adhesive tape. It consists of flat cells (corneocytes) containing keratin with lipid multilayers in between. Under "ordinary" humidity conditions, the stratum corneum contains about 10% (w/w) of water, whereas the water content in lower regions of the epidermis is about 70%. A healthy skin should be able to keep such a gradient even in very dry air.

Skin lipids are successively changed through the stratum corneum layer, and these changes have been followed in murine stratum corneum (Elias *et al.*, 1988). The lamellar bodies consist of glycosphingolipids and phospholipids. Phospholipases and sphingomyelinase digest these lipids, so that they become less polar. The predominant lipids in stratum corneum are ceramides (about half of the lipids), free fatty acids, free cholesterol and cholesterol esters. There is also a few per cent of polar lipids, dominated by cholesterol sulphate.

Electron microscopy and X-ray diffraction show that the lipids located in the intercellular space have a lamellar structure. There have been different proposals on the detailed arrangement, *e.g.* a 'trilaminar" unit layer (Elias, 1991). A lyotropic liquid-crystalline structure has also been described in extracted lipids from stratum corneum (Kayali *et al.*, 1991).

In a recent infrared (IR) spectroscopy study of porcine stratum corneum lipids, sheets of stratum corneum sealed between IR-transparent windows were examined at different temperatures by means of Fourier transform IR spectroscopy (Potts and Francoeur, 1993). According to changes in the C-H stretching vibration region, they found transitions in the temperature region 60-70°C corresponding to a melting of the hydrocarbon chains. These transition temperatures were confirmed by DSC-measurements. From their published IR-spectra, it is possible to obtain additional information on the organization of the hydrocarbon chains. The CH_2-rocking vibrations in the 720 cm^{-1} region reflect the hydrocarbon chain packing subcell (Chapman, 1962). The split of the peak in their spectra indicates that the chains exhibit the orthorhombic chain packing.

A unique feature of the skin lipid composition is the high content of ceramides. There are seven homologous series of ceramides, and the corresponding glucosylceramides are also present (Wertz and Downing, 1983). The ceramides are remarkable with regard to the lengths of the hydrocarbon chains; most of the fatty acids are in the range C_{24} to C_{30}. A further difference in the ceramide structure is the occurrence of hydroxyl groups in the fatty acid components, and occurrence of two different long-chain bases, sphingosine and phytosphingosine. Ceramides can form L_α-phases, but because of the long hydrocarbon chains, those in skin should be expected to form a gel-phase. This is confirmed by the IR-study described above, and by X-ray and thermal studies (Bowstra *et al.*, 1991). The strongly polar components, such as cholesterol sulphate and the small amount of phospholipids, should therefore be expected to segregate and form a separate aqueous bilayer phase. Solubilization of these polar lipids into weakly polar lipids can only take place if the chains have the liquid-type of disorder that is known from model systems (*cf.* Chapter 3).

A dual-compartment structure of stratum corneum has been proposed by Elias (*cf.* Elias *et al.*, 1983). A major factor behind the two-compartment model of the skin is the observation of separately located transport routes for polar and non-polar substances respectively through the skin barrier. This model has been described as a brick-mortar structure; the lipids form the mortar between the cells/bricks. A complementary model of the organization of stratum corneum lipids, which explains the barrier properties of stratum corneum in more detail, was recently proposed by Forslind (1994). It assumes crystalline bilayer units with bilayer regions with fluid chains in between ("domain mosaic model"). This model is able to adequately explain the penetration behaviour of hydrophobic as well as hydrophilic compounds.

Sebum is a non-polar oil, which spreads spontaneously on the surface of stratum corneum, as described above. This surface film is responsible for the hydrophobic surface characteristic of the skin. The function of sebum lipids and that of stratum corneum lipids have up till now been considered separately. Even if there are differences in location and physical properties between these groups, it is important to realise that they also interact. When these two lipid systems come into contact, there will be solubilization, molecular mixing and mutual effects on phase equilibria. A certain amount of the sebum will penetrate the stratum corneum and influence the state of the lipid structure of the intercellular space.

There is a wide spectrum of skin care formulations, ranging from solids (such as lipstick with pigment dispersed in a wax crystal matrix) to liquids (for example dilute o/w-emulsions for penetration of moisturizing agents). We will discuss the two extremes below. One is the formation of a layer outside the stratum corneum, without "chemically‘ influencing the skin (ignoring the effect on the water balance). The other represents the most common type of skin care product, and it involves penetration of active components through the skin, mainly in order to lubricate and moisturize the stratum corneum. We will simply call these types of formulations "external" and "penetrating' respectively.

B. SKIN CARE FORMULATIONS

1. External Formulations

Examples of formulations of this kind are pigment applications ("foundation"), sun-screening agents and skin protection creams. An efficient vehicle for these functions is provided by lipid crystal dispersions.

There are five different formulation types of lipids in skin care formulations: o/w- and w/o-emulsions, water formulations "immobilized" by hydrocolloids, oil formulations immobilized by lipid crystals and dispersions of hydrophilic lipid crystals in water. It is obvious that the water formulations can also be prepared free from lipids.

Cosmetic formulations can today be evaluated by scientific methods with regard to claimed effects. It will probably take a long time, however, before marketing of cosmetics will be based on such real knowledge. This is a matter of regret as there is a clear need for good cosmetic preparations, especially for the ageing skin. Some of the major cosmetic functions are:

- to increase the water content of the stratum corneum
- to provide environmental protection of the skin surface (against the sun, organic solvents *etc.*)
- to renew the barrier function by supplying lipid-like skin substances
- to improve lubricating properties between keratin units

The chemical substances used in cosmetic preparations should be selected with great care, as they can penetrate the skin and enter the circulation. Any substance administered on the skin will exhibit a partition between the applied film and epidermis. It will usually give some penetration and enter the circulation. Lipophilic and amphiphilic compounds penetrate easily, as discussed in connection with drug delivery (Chapter 13). The size of the molecules is of course another significant factor in penetration. For this reason, a good philosophy would be to use food components in cosmetics. It is remarkable that so many synthetic compounds, the degradative metabolism of which is not known, are still used in skin care products. It is likely that all desired cosmetic properties can be obtained by combining edible lipid components with hydrophilic compounds known to occur in the circulation and soft tissues (such as sugars, amino acids, salts, lactic acid or urea).

Hydrophilic lipid crystals dispersed in water were described in Chapter

7. Such a dispersion prepared from mixtures of monolaurin and monomyristin in about 70% of water (w/w) was observed to exhibit interesting properties when it was spread on the skin surface (Larsson *et al.*, 1975). Thus a flexible and invisible lipid film was formed on the skin after evaporation of water, with a structure like a shingle-roof. An electron micrograph of the dispersion is shown in Plate 11.1. The film allows perspiration and has good mechanical properties. The spreading on the skin surface involves opening up of the crystals at the methyl end group gap; the surface towards the hydrophobic skin surface regions, as well as the external surface, will therefore be hydrophobic.

This formulation has been used for two decades as a protective skin care formulation - an "invisible glove" - for protection against mineral oil (Larsson *et al.*, 1975). A significant reduction in mineral oil acne was found. Efficient protection against acrylate monomer penetration has also been observed (Glantz *et al.*, 1976). This kind of skin protection has been reported to be useful in eczema therapy (Klascha, 1982).

Another commercial application of this base is an antimicrobial formulation containing hydrogen peroxide (Glantz and Larsson, 1982). The lipids themselves are active against Gram-negative bacteria, and the hydrogen peroxide provides an additional broad-spectrum effect. Because of its location outside the skin (or a mucous layer), the hydrogen peroxide was found to keep its activity for several hours.

The crystalline type of external formulation has been described in detail here, as its structure is known down to the molecular level. The behaviour of other external formulations is somewhat more complex, as they contain lipids in disordered states. A certain degree of interaction with the lipids present naturally on the skin surface must therefore be expected. The other types of external formulations are oil-continuous or o/w-emulsions. After spreading, they will form a coherent oil-film covering the skin surface. The classical formulation of this type is a w/o-emulsion prepared from wool wax, as discussed in Chapter 8.

2. *Penetrating Formulations*

The keratin of the stratum corneum layer must contain a certain amount of water in order to possess good mechanical flexibility. The feeling of a "smooth" skin is therefore closely related to its water content. Especially in a cold and dry climate, when more water evaporates from the stratum corneum, drying of this layer can in the worst case lead to a scaly skin. There is also a need for lubrication in connection with all kinds of mechanical movement of the skin. The lipid layers of the intracellular space and the sebum lipids have good lubricating properties, but sometimes these are not sufficient. Frequent washing, for example, can remove too much of these lipids.

There are two different ways of increasing the water content, and both are related to physiological mechanisms for controlling skin surface evaporation. The first is the occlusion effect of the sebum film. If a formulation provides an oil film on the skin surface, less water will evaporate. Thus, the stratum corneum will maintain a higher water content. Cosmetic formulations have been based on this principle since ancient times, and lipids or lipid substitutes therefore represent the most common type of cosmetic ingredient.

Small polar molecules dissolved in water will penetrate the stratum corneum. The sweat is a natural source of compounds of this kind, such as salts and amino acids. These molecules provide an osmotic driving-force for water transport from the epidermis to stratum corneum. By adding glycerol, sugars, lactic acid and amino acids or urea, this effect is obtained. An additional effect of urea at a high concentration is solubilization of keratin.

A cosmetic preparation based on one or both of these principles of moisturizing the skin should not form a greasy layer on top of the stratum corneum. In order to give a thin and homogeneous film, the viscosity of the oil phase must be reduced. A substance used frequently for this purpose is isopropylmyristate. Wax oils, *i.e.* simple esters of long-chain alcohols and fatty acids, which are liquid at body temperature, have also the advantage of a low viscosity. One example is jojoba oil.

Only two decades ago, by means of measurements of water-binding to pieces of stratum corneum *in vitro*, a dermatologist showed that urea can moisturize skin. A cosmetic product, derived from this observation, was introduced and marketed. For many years, this was the only product in Sweden allowed to claim in advertisements that it could increase the water content of skin. This story can serve as an illustration of the lack of physiological knowledge that has characterized the field of cosmetics. It also indicates the possibilities for producing improved consumer products in this field based on new knowledge of lipids.

C. TRANSDERMAL PENETRATION

Drug delivery in general will be treated in Chapter 13. Drug penetration through the skin is closely related to the skin lipid organization, however, and it will be discussed in this context here.

The occlusion principle to enhance drug penetration through the skin is due to an increased water content in the stratum corneum. It can be achieved simply by applying a water-tight coating or by dressing the external skin surface, so that water saturation is obtained. Removal of skin lipids by washing with organic solvents or detergents is another way to increase penetration. This enhancement of penetration is much more pronounced when polar molecules are applied on the skin (Cooper, 1984). This is not surprising, as the barrier effect against polar molecules is the hydrocarbon chain core of the lipid bilayers. If they are partly removed, there will be openings. Lipophilic molecules on the other hand need lipid bilayers for their movement through the skin. Partial removal of the bilayer will not have much effect.

Certain amphiphilic molecules can strongly enhance the penetration of lipophilic drugs. Ethyl acetate has been shown to increase the penetration of levenorgestrel, a contraceptive agent (Friend and Heller, 1993). Azone (1-dodecylazacycloheptan-2-one) is a substance developed in order to enhance skin penetration (Hadgraft *et al.*, 1993), and there are numerous reports of its effect on transdermal drug transport. The interaction between azone and lipid-water systems as models of skin lipids has recently been reported (Engblom and Engström, 1994). It was shown that the incorporation of azone into lipid bilayers will induce a phase transition from the L_α-conformation into a cubic bilayer structure. The penetration enhancement was therefore proposed to be due to the formation of a bicontinuous structure in the intercellular space of the stratum corneum.

There is a pH-gradient from about 4.2 - 5.6 on the skin surface to 7.3 - 7.4 on the dermis side (Katz, 1973). This gradient can be used to provide energy for the transport of anionic drugs. Amines can complex with anionic drugs, and in this way facilitate penetration (Barker and Hadgraft, 1981). Further examples of penetration enhancement of drugs due to interaction with skin lipids will be given in Chapter 13.

It should finally be mentioned that the natural surface protection by a lipid coating of skin, discussed here in the case of humans, occurs also in other animals and in plants. Insects, for example, are effectively protected against dehydration by *n*-paraffins in the α-crystalline state. The green parts of all plants are covered by *cutin*, a lipid-polymer network which becomes water-proofed by an oil film of small lipid molecules. A most spectacular effect of lipid surface films is illustrated by seabirds. Highly branched fatty acids esterified with alcohols form the preen gland wax, and a film of this oil over the plumage protects the bird against water penetration. Formulations based on such branched lipids have been reported that can help the bird to repair this protective film, after it has been destroyed by mineral oil following oil pollution accidents (Larsson and Odham, 1970).

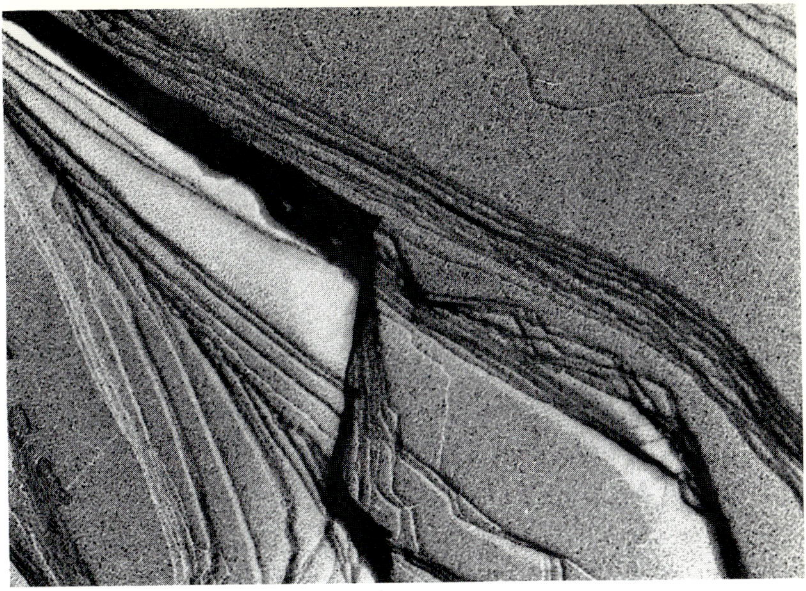

Plate 11.1. An electron micrograph of hydrophilic monoglyceride crystals prepared as an aqueous dispersion. The smallest fracture steps correspond to the lipid bilayer thickness. The micrograph was kindly supplied by Dr. A.-M. Hermansson.

CHAPTER 12

POLAR LIPIDS IN GASTROINTESTINAL AND PULMONARY SURFACE LAYERS AND ADMINISTRATION OF EXOGENOUS PHOSPHOLIPIDS

A. MUCOUS LAYERS AND MEMBRANE LIPIDS

The skin, our protection against the outer environment, is a dry surface characterized by an outer film of non-polar lipids. We also have inner surfaces that are wet. They consist of hydrophilic proteoglucans, mucins, which form an aqueous gel-like phase, mucous. A surface film of polar lipids is often a part of the mucous zone, with a major influence on the surface properties. One common function of the flow of a hydrophobic film on the surface of the skin (sebum) and the flow of hydrophilic mucous above inner surfaces is to keep the underlying biological tissue clean and intact.

Below the mucous layer, there is an endothelial plasma membrane, which is successively renewed. The transport of membrane lipids into the lumen via the mucous layer induces amphiphilic properties at mucous surfaces, a turnover which has often been neglected.

A phospholipid, or "surfactant", monolayer is crucial for lung function. Without the presence of a condensed phospholipid monolayer, the lung alveols would collapse. Gas exchange is also enhanced by the lipid surface film. The lung mucosa represent an extreme where the presence of a lipid monolayer at the air/water interface has a fundamental physiological function. Another extreme can be represented by saliva and the oral mucosa, where there is no lipid monolayer covering the surface. This is evident from surface tension studies of saliva (Sefton *et al.*, 1992). The reduction in surface tension at the air/water interface by the saliva components is about 30 mN/m, which is clearly lower than the value expected if polar lipids were to form the interface. There have also been analytical studies of lipids in saliva (Slomiany *et al.*, 1985), demonstrating extremely low levels under healthy conditions. Sphingolipids, glyceroglucolipids and glycerophospholipids were found. About 90% of the saliva is secreted from parotid and submandibular glands. This saliva is dominated by proteins and contains only about 10 mg/100 ml of lipids. The proteins in the saliva are strongly surface-active, forming a protective interfacial film in the whole oral cavity.

B. LIPIDS IN THE GASTROINTESTINAL BARRIER

1. The Gastric Mucous Surface

The gastric wall contains surface mucous cells, secreting mucous and phospholipids. Before secretion, the phospholipids are stored as "lamellar bodies", which resemble liposomes. Mucous consists of mucin molecules, large proteoglucans. Bhaskar *et al.* (1992) have recently demonstrated that hydrochloric acid exposed to gastric mucous forms "viscous fingering patterns". In an elegant study, they were able to show that hydrochloric acid is prevented from diffusing back to the epithelial surface by the high viscosity of "a gelled mucous boundary". Their description of the physical properties is fully consistent with a two-phase separation. It is therefore proposed here that the gastric mucous layer is a two-phase aqueous system in a dynamic non-equilibrium state. The evidence is partly based on the effect of endogenous lipids on the wetting behaviour discussed below.

If an aqueous protein solution is successively concentrated, a separation into a protein-rich and a water-rich phase will usually take place (Bungenberg de Jong, 1949). The protein-rich phase is described as "coacervate" in this kind of two-phase separation phenomenon. In the gastric mucous layer, the proteoglucan (mucin) molecules are constantly secreted into the coacervate layer, and the water-rich layer forming the outer phase is constantly diluted due to the luminar flow. Equilibrium between the phases is never reached, as mucin and water molecules pass through the interface between the two phases.

The occurrence of two aqueous phases, which we will here call the mucous phase and the luminar phase, has important functional effects. The partition of a molecule between the two phases is related to its molecular weight, and with increasing molecular weight the partition coefficient will approach a value of unity. A particle will therefore choose one phase or the other, depending upon its surface properties. Alternatively, it will avoid both phases and form a third phase. Because of this property, the mucous phase will function as a filter, which can hinder certain macromolecules and microorganisms from passing the mucous layer.

It is likely that the partition of phospholipid aggregates, such as those secreted from surface mucous cells, in the choice between a water-rich and a proteoglucan-rich phase will go to the water-rich phase or form a third phase at the interface between the two phases. Considering the phospholipid aggregates known to be secreted from the gastric mucosal cells, this partition means that they will be squeezed out from the mucous layer directly into the luminal phase. This is confirmed by ultrastructure studies discussed below.

Davenport (1970) reported that concentrations as low as 100 μg/ml of lysophosphatidylcholine can damage the gastric mucosal barrier of dogs, and indications that lyso-PC is a risk factor for gastric ulcer in humans have also been reported (Johnson and McDermott, 1974). Numerous recent papers claim that the lumen surface of the gastric wall is hydrophobic because of the presence of phospholipids, which provide cytoprotection (*cf.* Swarm *et al.*, 1987). In order to obtain some information on the behaviour of PC under conditions corresponding to those in the gastric lumen, some simple experiments were performed as described below.

When the pH of an aqueous PC dispersion is lowered, the phosphate groups of the molecules will successively become undissociated. The pK-value of PC bilayers in water is related to the degree of hydration and the detailed packing in the polar region of the bilayer, and a reported value of 1.5 is therefore not generally valid (Israelachvili, 1985). In order to estimate the degree of dissociation at the gastric pH, 5% (w/w) of PC was dispersed in hydrochloric acid with a pH = 1. There was only a shift in pH of 0.05, and titration to pH = 1 showed that only about 25% of the PC molecules occurred as the hydrochloric acid salt. Because of hydrolysis of PC (described below), it was not possible to determine the pK-value accurately by titration. It remains evident that a large proportion of the PC in the gastric lumen is cationic, however.

The TLC pattern of aqueous phases of PC with the pH shifted to correspond to that in the gastric environment shows that after 5 hours, a significant degree of hydrolysis has taken place at pH = 1. Even at pH = 2, a small proportion of PC has been hydrolysed.

When the zwitter-ionic phospholipid becomes partly cationic, the phase behaviour is also changed. We saw in Chapter 3 that PC takes up water at a neutral pH into a lamellar liquid-crystalline phase, with a limit of swelling of about 40% (w/w) of water. Thus a PC dispersion in the presence of water in excess forms liposomes at neutral pH (easy to identify in the polarizing microscope). When the pH was shifted to a value of 1, however, a 5% (w/w) PC dispersion in water formed a homogeneous and weakly birefringent liquid-crystalline phase. Thus the presence of net charges in the bilayer can be detected as the lamellar phase then shows "infinite" swelling due to electrostatic repulsion. The presence of sodium chloride at a physiological concentration did not alter this swelling behaviour. However, a remarkable observation was the presence of some strongly birefringent liposomes, indicating that there is a segregation of bilayer aggregates with zwitter-ionic PC molecules only.

In order to get some indication of whether the gastric mucous layer provides a protective effect against PC hydrolysis, we tried to simulate the mucins by using an agar solution. The mucins consist mainly of peptide backbones with sugar group branches, whereas agar is a linear polyanion. Electrostatic complexing and shielding of PC bilayers would be expected to be similar.

With PC dispersions of neutral pH, the agar solution gave just a thickening effect, and the liposomes showed the same appearance in the polarizing microscope as in the absence of agar. It can therefore be assumed that the agar molecules are distributed in the "free" water phase outside the liposomes. When the pH was shifted to a value of 1, a precipitate was formed when agar (also at the same pH) was added to the PC-water phase. The mechanism must be electrostatic complexing.

When PC hydrolysis was followed with agar present, it was not possible to detect any difference from when agar was absent. This result indicates that mucins cannot protect PC against hydrolysis induced by low pH. Now returning to Davenport's results on lyso-PC, the flow of PC from the surface zone of the gastric wall into the intestinal system must be rapid enough to avoid hydrolysis of PC. Lyso-PCs like detergents form micellar solutions, which can also solubilize cytoprotective phospholipids as discussed below.

The secretion of phospholipids from the gastric mucosal cells was mentioned above (*cf.* Kao and Lichtenberger, 1987). Their protective role, for example in mucosal damage induced by aspirin, has also been demonstrated (*cf.* Swann *et al.*, 1987). An oligolamellar lining, consisting of phospholipids, has been demonstrated at the gastric wall, and it has even been proposed to constitute the physical identity of the gastric barrier (Hills, 1990). Numerous recent studies based on contact angle measurements have been interpreted as a demonstration of a "hydrophobic lining" of the gastric surface, and relationships between reduction in this hydrophobicity and peptic ulcer disease have been reported (Goggin *et al.*, 1991). In the literature on the hydrophobic lining of the gastric wall, there is a basic misunderstanding, however. The surface is thus considered to be hydrophobic under physiological conditions as a consequence of the orientation of the phospholipid molecules, and this is not correct. The measurements merely reflect the amphiphilic properties induced by the presence of phospholipids at an interface between the mucous phase and the luminar phase. This will be illustrated below by a simple model experiment.

When a liposomal dispersion of a few per cent of phosphatidylcholine, for example egg yolk PC, is kept in a glass vessel, the water phase does not wet the glass wall. Water droplets roll over the surface, like droplets of mercury. In other words, the hydrophilic glass surface appears to be hydrophobic. Between the glass surface and the water there is either nothing or bilayer units of PC, whereas at the air/water interface there is a condensed monolayer of PC. Let us now consider the interface between

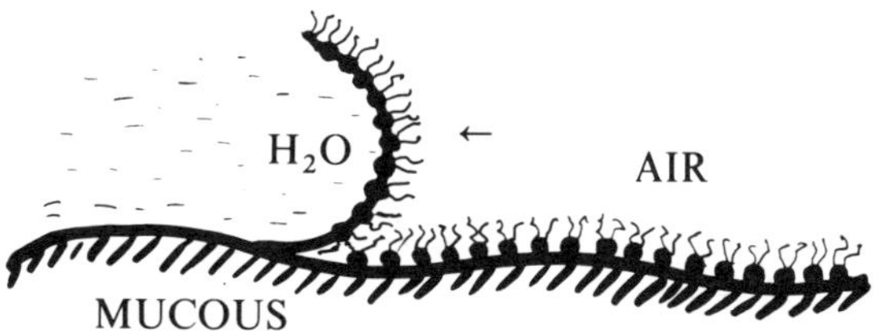

Fig. 12.1. Schematic illustration of surface monolayer formation when the gastric luminar phase in contact with the mucous phase is replaced by air prior to contact angle measurements.

the gastric mucous phase and the luminar phase when a piece of the gastric wall is taken out for contact angle measurements. When the liquid luminar phase leaves the mucous phase, condensed monolayers will cover the surfaces exposed towards air in order to reduce the surface free energy, as indicated in Figure 12.1. Cationic PC molecules interacting with negative groups in the mucin molecules might strengthen the adherence of the PC monolayer.

The deposition of a water droplet on this mucous phase, covered by a condensed PC monolayer, will exhibit wetting behaviour corresponding to that of a hydrophobic surface. When water covers the surfaces, however, this monolayer will associate into bilayer units; the surface can thus switch between hydrophobic and hydrophilic depending upon the environment.

When there are not enough PC molecules to give a condensed monolayer, the surface will appear hydrophilic, whereas the presence of enough phospholipid molecules will indicate a non-wettable surface. It is obvious that this situation should be expected to change following oral administration of phospholipids, as has been observed (Hills, 1990). An exogenous supply of PC has recently been shown to provide protection against gastric ulcer formation (Dunjic *et al.*, 1993).

Finally it must be mentioned that a gastric lipase has been found, which with a lipase in the saliva, contributes to the digestion of dietary fats in the stomach (Gargouri *et al.*, 1989). These enzymes do not hydrolyse phospholipids significantly. A surface film of PC molecules covering the mucous phase should not therefore be influenced by such lipases, whereas the hydrolysis due to the low pH means that the interfacial PC is replaced within about an hour. Two common substances are known to injure and penetrate the gastric wall, aspirin and ethyl alcohol. It seems likely that they act by modifying the two-phase separation described above.

2. *The Surface Layer in the Intestines*

Mucous is secreted along the whole intestinal wall. The small intestine is particularly efficient in digesting and absorbing food nutrients, whereas water absorption is a major function of the colon. Common features of this mucous region are the presence of microorganisms and the location of a significant part of the immune system. T-Lymphocytes in the epithelial cells control reactions against microorganisms and food components (together with their digestion products).

The low gastric pH provides an effective protection against microorganisms. When food enters the small intestine, the pH-value is normalized. An antimicrobial protection in this region is provided by "friendly" microorganisms, which inhabit the mucous layer. Lactobacilli inhibit the growth of potentially pathogenic bacteria (Johansson *et al.*, 1993), and they are even considered to stimulate the immune response (Fuller, 1991). They are sometimes described as probiotics, therefore. Their packing density at the surface increases successively along the intestinal system, from about 10^6 *CFU* (colony-forming units) per ml mucosa in the small intestine to about 10^8 *CFU*/ml in the colon. Colonization with different Lactobacillus strains has been followed *in vivo*, and a corresponding decrease in gram-negative bacteria was observed (Johansson *et al.*, 1993). Among significant bacteria, *Helicobacter pylori* should be mentioned. In recent years, it has been demonstrated that it is often involved in gastrointestinal inflammatory diseases and ulcer formation. These bacteria can neutralize the low gastric pH through extracellular urease activity. By means of their flagella, they can move through the mucous layer, and they can then adhere to glycerolipid receptors at the gastric wall (Lingwood *et al.*, 1989).

Colonization of the intestinal epithelial surfaces by Lactobacilli is host specific (Chagnaud *et al.*, 1992), so that a strain that can colonize the mucosa in one animal species will not normally colonize the epithelium of another. The molecular mechanisms responsible for the adherence of Lactobacilli are not known. Efforts to relate the cell surface proteins of Lactobacilli to adhesion have up till now been unsuccessful (*cf.* Chagnaud *et al.*, 1992).

The surface area of the intestine is comparable to that of a football ground, and the cell membrane is renewed every second day. This means that large amounts of membrane lipids become a part of the mucous layer. The mucous-surface also in this region is considered to be "nonwettable" because of the presence of phospholipids (*cf.* Butler *et al.*, 1983). It is not known at present whether the apparent hydrophobicity is related to a two-phase separation, as proposed above in the case of the gastric wall. The brush border of the small intestine is covered by a "fuzzy coat‘ constituted by the sugar chains attached to lipids and proteins of the epithelial cell membranes. Outside this layer, there is a film of mucous, which constantly flows from the mucosal cells. The "fuzzy coat" or glycocalyx might be regarded as a separate phase in relation to mucous.

There is a high proportion of glucolipids among the lipids in the outer surface of the intestinal membrane. When these contain up to ten or more sugar groups, their solubility must be considerable, as discussed in

Chapter 9 (Section D). The solubility aspects of the most polar membrane lipids have been neglected. It seems possible that a loss of glycolipids into mucous, because of solubility, is an important mechanism behind renewal of the intestinal membrane.

The lipid spectrum in the intestinal mucosa is quite complex. In addition to the lipids of the epithelial membrane, we must consider the bile (containing phospholipids, bile acids and cholesterol mainly) and food lipids with their degradation products, produced by the pancreatic enzymes.

A special group of lipids plays an important role in the mucous layer of colon, *i.e.* short-chain fatty acids (*SCFA*s). They are produced by bacterial fermentation of dietary fibre. It was not realized until rather recently that *SCFA*s, produced in the lumen and mucous surface layer of the colon, are the primary energy supply for the endothelial cells of this organ (Roediger, 1981). The absorption of *SCFA*s also seems to have a regulatory function in the hepatic metabolism of carbohydrates (Anderson and Budges, 1982). The absorption mechanism is considered to be passive diffusion (Hoverstad, 1986). The penetration of the membrane of colonocytes might involve the type of lipid bilayer phase changes occurring in the related phase diagram of *n*-butanol/PC/water, described in Chapter 3. We have found that butyric acid and butanol have similar effects on PC-water phase equilibria.

Administration of phospholipids exogenously has been shown to prevent acetic acid-induced colitis (Fabia *et al.*, 1992). In the study performed on rats, PI gave a better effect than PC, although both showed a good therapeutic effect.

3. Intestinal Absorption of Lipids

Triglycerides are digested by pancreatic lipase/colipase and carboxylic ester lipase. There is an interplay between these two enzymes in the absorption of the nutritionally important C20:4, C20:5 and C22:6 fatty acids (Chen *et al.*, 1989). Other fats can be digested into fatty acids and 2-monoglycerides by lipase/colipase only. The partial digestion of dietary lipids by lingual and gastric lipases is ignored in this connection. The phase behaviour of the lipolysis products is one significant factor in the adsorption process (*cf.* Lindström *et al.*, 1981). Bile acids contribute to micellar solubilization of monoglycerides and fatty acids next to the brush-border membrane, and the monomers are taken up by specific proteins in the membranes of those cells involved in lipid absorption, the enterocytes (Stremmel, 1988).

Phospholipids in the diet, from the bile, or released from the brush-border membranes are digested by pancreatic phospholipase A_2, and the monomers enter the enterocyte membrane. In the enterocytes, re-esterification takes place into triglycerides via the monoglyceride pathway and into phospholipids via the phosphatidic acid pathway. Pre-chylomicron particles are then formed and exported into the lymphatic system. In the case of "medium-chain" fatty acids (C8-C10), the free acids are directly exported into the portal blood.

C. THE PERITONEAL MESOTHELIUM

Examination of the peritoneal effluent obtained during peritoneal dialysis has shown the presence of phospholipids, and it was proposed that their function is to repel water and lubricate the tissue (Grahame *et al.*, 1985). Furthermore, a decrease in the concentration of phospholipids has been observed during long periods of dialysis (De Paolo *et al.*, 1986). The frequent occurrence of tissue-adhesion formation after peritoneal dialysis has been related to the loss of phospholipids from the surface of mesothelial cells. It was thus natural to assume that surface layers of phospholipids prevent adhesion, and this has been confirmed experimentally (Ar'Rajab *et al.*, 1991). Intraperitoneal adhesions were induced in rats, and it was shown that the administration of PC into the peritoneum could prevent these.

The occurrence of a surface layer of phospholipids, mainly PC, in the peritoneum seems to be well documented. The phospholipids can hardly form a hydrophobic surface in the aqueous environment in the physiological situation, as has been proposed (Grahame *et al.*, 1985). The arguments are the same as those given above in the discussion of the gastric wall.

D. THE PULMONARY AIR/WATER INTERFACE

In the introduction to this chapter, we discussed some general features of the surface layer in the lungs. The term "lung surfactant", which is used in the literature, is perhaps not adequate. Surfactants are water-soluble amphiphiles, as described in Chapter 3. In connection with the lung tissue, however, the term is used for the whole monolayer complex, including proteins. Certain alveolar epithelial cells (type II) synthesise phospholipids which are stored as lamellar bodies and released by exocytosis, providing a surface layer of "surfactant" in the lungs. The pulmonary surfactant consists of about 80% (w/w) lipids and 10% proteins (Harwood *et al.*, 1986). Two proteins (molecular weight about 10,000 and 35,000 Daltons) predominate and are considered to have specific functions (Harwood *et al.*, 1986).

The lamellar bodies, after secretion from the type II cells, swell and form "tubular" structures at the alveolar surface serving as depots for the phospholipid interface. These depots were described as myelin figures, but according to later freeze-etching electron microscopy their structure was defined as fine parallel grooves (Untersee *et al.*, 1971). It therefore seems likely that the structure must be closely related to the cubic or L3 lipid-water phases (see Chapter 3).

The surface tension at the air/water interface in the lungs reflects the surface structure, and its changes with interfacial area determine the energy required for breathing. Hill (1988) has recently presented interesting new ideas on alveolar functions, and in this connection reviewed the surface tension and the equilibrium spreading pressure of the lung surfactant system. There are two schools; one reports values close to zero (0.01 mN/m) and the other finds values of about 20 mN/m. The higher values are close to the ESP-value of dipalmitoyl-PC (DPPC), which is the predominant component of this surfactant mixture. It is interesting to note that DPPC monolayers show a remarkable hysteresis behaviour on repeated compression-expansion cycles (Hills, 1988). During the first cycles, the hysteresis corresponds to a higher surface tension on expansion, but after about 20 cycles, the hysteresis is inverted, indicating that mechanical work is performed by the monolayer. The origin of this energy is not known. A related rhythmic pulsation of bubbles from excised lungs in saline solution has been observed. The hysteresis represents a two-dimensional engine which could reduce the work of breathing by about 23%.

The explanation of the hysteresis of DPPC might be related to the occurrence of the rippled bilayer phase of this lipid, discussed in Chapter 2. We learned in Chapter 5 that monolayers of polar lipids have a structure which is directly related to the corresponding lipid-water phases in bulk. At physiological temperatures, the rippled phase exists in DPPC in the presence of excess of water, and it should therefore be expected to exhibit a monolayer analogue at the air/water interface. Furthermore, such an "odd" monolayer structure might require some kind of heterogeneous nucleation initiated by the repeated compression-expansion cycles. During the first compression-expansion cycles, a "normal" condensed monolayer with crystalline chains is proposed to occur. Later the rippled structure is formed. The phase transition into a rippled monolayer might involve extra thermal energy, later released as mechanical work.

The full physiological significance of the surfactant layer is still a matter of controversy, and various alveolar models have been proposed (*cf.* Hills, 1988). A basic difference between "dry" and "wet" models concerns the subphase under the surface monolayer, *i.e.* whether there are aqueous regions or not. The occurrence of an L3-phase instead of water might explain the controversy. (The ultrastructural evidence for an L3- or cubic structure was mentioned at the beginning of this section).

Hills (1988) has reported convincing data, demonstrating that the alveolar surface consists of a few lipid bilayers, with a monolayer forming the air/water interface. This monolayer is responsible for the "dry" hydrophobic surface properties of alveols.

Premature infants can die as a consequence of a lack of lung surfactant at their birth; too much energy is required to expand the lungs, and on expiration the lungs collapse (*cf.* Hallman *et al.*, 1991). This condition is called *respiratory distress syndrome.* Tests for lung maturity should involve both the concentration of PC in the amniotic fluid and the surface tension or foam stabilization (Harwood *et al.*, 1986). Promising trials in treating this syndrome with phospholipid preparations, with synthetic surfactants, or with lung surfactant from animals have been reported. On the other hand, no therapy has yet been granted clinical acceptance, and this may be related in part to the uncertainties regarding the functions of lung surfactant, discussed above.

CHAPTER 13

DRUG DELIVERY

Most drugs are solids, which are taken orally. When they are formulated into pills from powders, lipids are used as solid lubricants. This chapter is concerned with other formulations where the liquid or liquid-crystalline structure of the lipid phases plays a direct role in drug delivery.

Drugs are usually amphiphilic with limited water solubility. In the case of lipophilic drugs, the addition of alcohols can increase the solubility (ethylene- or propyleneglycol and ethyl alcohol are used). However, a superior strategy is to introduce lipids in the formulation when designing effective drug-carrying systems. By using the whole spectrum from non-polar lipids to the most polar ones, going successively from oil-continuous via bicontinuous to water-continuous formulations, it is possible to solubilize almost any type of drug. An alternative technology for drug delivery is micro-encapsulation. The first application was gelatin coating of particles of the drug by a so-called coacervation process, which involves two-phase separation of a water solution (see Chapter 12). One phase will be rich in gelatin, and the particles are dispersed in this phase. At present the use of biodegradable polymers for production of microspheres ($\langle$2500 Å) is a very active field of research. The polymers can be starch or polyesters based on lactic acid or glycolic acid. Such formulations are used for parenteral drug delivery. In the following discussion, we will only consider lipid formulations.

By utilizing the special features of lipid-water structures, there are also possibilities to stabilize drugs, to mediate their absorption through skin or mucous layers, and in certain cases to achieve targeting. The different types of lipid dispersions or phases of interest will be discussed below.

Some general comments on drug penetration through membranes were given in Chapter 9. If we consider the special case of drugs with a site of action in the skin or within a mucous layer, penetration is not desired. When lipids are used in such carrier systems, they should not disturb the cell membrane barrier but stabilize it. Strongly polar surfactant-like lipids should not be used therefore, whereas non-ionic or zwitter-ionic membrane lipids can be expected to stabilize the membranes against penetration.

Water-in-oil emulsions have been used for intramuscular injection of drug formulations, for example for controlled release of anticancer drugs (Hashida *et al.*, 1977). After injection, the drug will successively reach the lymphatic system and the bloodstream. Multiple emulsions (w/o/w) have been described for the delivery of methotrexate (Elson *et al.*, 1970).

2. *Oral Emulsions*

Fat soluble vitamins (A, D and E) are usually delivered in emulsion formulations, and it should be pointed out that such emulsions occur "naturally" in foods.

The oral absorption of drugs is often limited if they have low solubility in water; griseofulvin can serve as an example. If such drugs can be solubilized in an oil, an appropriate emulsion can enhance the absorption. The administration of griseofulvin in corn oil has demonstrated this effect (Carrigan and Bates, 1973).

Oral absorption of heparin has been achieved by administration in corn oil emulsions stabilized by sodium taurocholate (Engel and Fahrenbach, 1968). The mechanism is not known. When the oil phase is digested enzymatically, there is a complex series of events, transforming the emulsion droplet into liquid-crystalline and micellar phases (*cf.* Lindström *et al.*, 1981). If a drug is solubilized into the oil, the partition of drug molecules between the outside water phase and these different phases will influence the absorption process. It is therefore not possible to predict the degree of absorption in general terms; model studies of the partition between the phases are needed.

3. *Administration of Emulsions on the Skin*

Drugs for transdermal delivery can be dissolved in the oil or water phases of an emulsion, and o/w- as well as w/o-emulsions are used. A general feature is an increased penetration in the case of w/o-emulsions, that has been assigned to the so-called occlusion effect (as described in Chapter 12). This means that the water content of stratum corneum is increased because of reduced water evaporation, and the barrier function is then reduced.

C. THE L_α-PHASE AND LIPOSOMES

If a drug is solubilized in the L_α-phase, it will tend to form a liposomal dispersion in contact with an aqueous medium. A film of the L_α-phase spread on the skin will keep its bilayer structure until the lipids are taken up by the skin lipids. An L_α-phase taken orally will, however, form a liposomal dispersion spontaneously. An enormous amount of work has been devoted to the application of liposomes for drug delivery. Liposomes and vesicles are described in Chapter 7, and their possibilities for the incorporation and transport of drug molecules are discussed below.

The laboratory technique used most often to prepare liposomes is simply shaking or stirring. More energy input is needed to get a uniform vesicle dispersion, and ultrasonification can then be used. More sophisticated technology, particularly for production of giant vesicles, has been introduced. Liposomes are then transformed into vesicles by extrusion (Olson *et al.*, 1978) or by dialysis (Szoka and Papahadjopoulis, 1978). So-called reversed phase technology (using organic solvents) is another alternative (Hope *et al.*, 1986). The development of processes for industrial scale production has also been reported (Kikuchi *et al.*, 1991).

A short and good review of the application of liposomes as drug carriers was recently published (Annesini *et al.*, 1992). If liposomes or vesicles are used for encapsulation, their permeability is a rather significant factor. The following examples of permeability through a single phospholipid bilayer provide some guidance in evaluating application potential (*cf.* Annesini *et al.*, 1992):

- water	10^{-2} cm/s
- urea, ethanol	10^{-5} - 10^{-6} cm/s
- glucose	10^{-9} cm/s
- Cl^-	10^{-10} cm/s
- Na^+	10^{-12} cm/s

The permeability coefficient (cm/sec) gives the flow in moles per second through a one cm^2 membrane according to the concentration difference between both sides.

Non-ionic molecules follow the permeability expected from solution-diffusion theory rather closely. The permeability of ions is, however, much higher than expected, especially the transport of protons and hydroxyl ions. There is no model at present to account for this.

The mobility of lipids from one side of the bilayer to the other ("flip-flop") is another important factor when certain lipids or lipid-analogues constitute the active component to be delivered. The movement of free fatty acids from one side of the bilayer to the other in phospholipid vesicles was recently determined in a very elegant experiment (Kamp and Hamilton, 1992). They used a fluorescent probe, which was pH-sensitive, and this probe was entrapped inside phosphatidylcholine or phosphatidylcholine/cholesterol vesicles. Free fatty acid addition from the outside then resulted in an inside pH-shift due to flip-flop, which could be seen directly by the pH-probe trapped inside. It was concluded from this work that unionized fatty acids flip-flop in the time scale 1.5 to 2 seconds, whereas ionized fatty acids flip-flop in a time scale of minutes.

When vesicles or liposomes are used as drug carriers, a hydrophilic (non-amphiphilic) drug must be entrapped in the water compartment, whereas lipophilic or amphiphilic drugs can be solubilized in the lipid bilayer. There are numerous technical problems with applications of the first alternative; it is hard to achieve a good yield in the entrapment process, and the drug will leak out during storage. The use of lipophilic drugs is the ideal application, particularly when the vesicle is taken up completely by the target-cell by endocytosis. Therefore, this approach is often tried even in the case of water-soluble drugs by first making the drug lipophilic through linking it covalently with a hydrocarbon chain region.

Vesicles or liposomes have a very short life-time usually when introduced into the circulation. The rapid clearance from the blood is due to uptake by the reticulo-endothelial system (RES) of the liver and the spleen. By applying simple principles of colloidal stability, their time of circulation has, however, been prolonged considerably. The use of lipids with many sugar groups (gangliosides) or synthetically modified lipids with polyoxyethylene groups can achieve steric stabilization (as discussed in the emulsion chapter). The expression "stealth-liposomes" has been coined as a description of such liposomes, which can be introduced into the circulation without being detected by the RES.

An example of the use of vesicles in the bloodstream is the replacement of erythrocytes as oxygen-carrier. This was achieved by encapsulation of haemoglobin, and a circulation half-life of about 12 hours was obtained (Beissinger *et al.*, 1986).

Although considerable work has been invested in the liposome/vesicle approach to drug delivery during the last two decades, no applications have yet reached the market, with a few exceptions in dermatology. The concept is still considered to be very promising though, and it is probably only a matter of time until various applications are implemented. Some of these drug formulations and some of the obstacles will be summarized.

Encapsulation of antimicrobial and anticancer agents has been studied successfully in animal models and is now in a clinical trial, for example for delivery of doxorubicin (Juliano and Stamp, 1978). Another promising field for liposomes is immunology. Liposomes are tested clinically for delivery of the immunomodulator muramyldipeptide, and an injectable liposome-based vaccine against malaria has been tested in humans (Alving and Richards, 1990). The main reason for using liposomes is their adjuvant effect; the immunogenic response of a particular peptide is thus strongly amplified by its location in the vesicle. This is beyond doubt a very important future field of applications.

The entrapment of enzymes to treat patients suffering from enzyme deficiency diseases is an interesting application of vesicles and liposomes. Enzyme activity can be strongly enhanced, because of the change in microenvironment, when the enzyme is located in a vesicle. As an example it has been shown that prostaglandin synthetase is increased in activity by liposome entrapment (Sada *et al.*, 1987).

It can finally be mentioned that covalent cross-linking has been introduced to stabilize liposomes and vesicles. This is certainly of interest in studies of cell membrane functions, but because of toxicology uncertainties it is of less interest than drug carrying systems.

D. H_{II}-PHASES, CUBIC PHASES AND CUBOSOMES

The H_{II}- and cubic phases can coexist in equilibrium with excess of water. Hydrophilic drug molecules can be solubilized in the aqueous compartment of these phases, whereas lipophilic and amphiphilic drug molecules will be located in the lipid bilayer. A wide range of different types of molecules can therefore be carried in the H_{II}- and cubic phases, and as described in Chapter 3 it is even possible to incorporate macromolecules, such as proteins. On this basis, the application of the H_{II} and cubic phases as general drug carriers has been developed (Engström *et al.*, 1992). The first commercial application is a controlled-release formulation of an antibiotic, metronidazole, for administration in the pockets between tooth and gingiva to act against periodontitis (Norling *et al.*, 1992). The antibiotic is dispersed in a liquid (L2) lipid phase, which can be injected easily into the subgingival pockets, where it swells and forms an H_{II}-phase. The pathogenic bacteria in the pocket are killed whereas the oral flora is not affected, because of the controlled-release mechanism of this formulation, Elysol™.

The most promising application of cubic lipid-water phases in drug delivery involves peptide drugs. As shown in Chapter 7, cubic lipid-water phases can incorporate proteins and even stabilize their native structures. When the monoolein-water cubic phase, for example, is injected into soft-tissues, it is metabolized by lipase activity, and solubilized peptides are released.

A disadvantage with cubic and H_{II}-phases is their three-dimensional texture. A drug delivery system for general use should consist of particles, since only particle dispersions can be injected into the blood-stream or administered into the lungs. The mechanism for the formation of cubosomes, as described in Chapter 7, provides such a possibility. One condition to form cubosomes in a system consisting of lipid-water phases and water is location in a three-phase region, cubic phase + L_α-phase + water (Larsson, 1989). An alternative is to disperse the cubic phase by adding an amphiphilic polymer. Landh (1994) has demonstrated how stable cubosomes with wide variations in structure can be produced in this way. On this basis, it is apparent that cubosomes will have possibilities for applications comparable to those of liposomes in the future. A typical cubosomal dispersion is shown in Plate 13.1.

The differences between cubosomes and vesicles/liposomes should be taken into account when applications are considered. If a large water compartment is needed, for example in order to provide high water activity of an entrapped enzyme, vesicles or liposomes should be chosen. If a large lipid bilayer area is desired per unit volume, cubosomes represent the optimal solution. Also, if proteins in the water medium are stabilized when surrounded by a minimum amount of water (a situation described in Chapter 3), cubic phases are the best solution.

The limitations in our present possibilities to apply vesicles/liposomes or cubosomes for injection into the bloodstream are a lack of knowledge of their interactions with blood components and therefore a lack of means to control targeting. How do these two types of artificial membrane assemblies interact with the different lipoproteins, how will the entrapped drug enter the target cell, and how much of the drug will get access to other tissues? On the technology side, more knowledge is needed in order to produce the desired surface properties of these aggregates for targeting.

Cubosomes, like liposomes, can be used as antigen carriers with applications as immunoadjuvants for vaccines. Thus, the particulate character means that they induce both a cellular-mediated immune response and a humoral immune response. Cubosomes covered by pluronics appear to be protected for discovery by RES, and they might therefore be termed stealth cubosomes by analogy with stealth liposomes.

E. SURFACE-ACTIVE LIPID CRYSTALS

The possibility to form hydrophilic lipid crystal dispersions in water was described in Chapter 7. Further, we saw in Chapter 8 that polar lipid crystals can be prepared so that they expose a hydrophobic side towards oil and a hydrophilic side towards water.

The hydrophilic lipid crystal dispersion of monolaurin and monomyristin in water has been found to stabilize hydrogen peroxide. An antimicrobial formulation based on this mechanism has been developed for application in wound healing (LHP™, produced by Hydro Pharma, Malmö, Sweden). It can also be applied on mucous layers (Glanz and Larsson, 1982). The release of monomers of monoglyceride molecules from the crystals provides an additional antimicrobial effect of this formulation.

Dithranol is an effective agent against psoriasis, but the therapy has been complicated because of oxidative degradation of dithranol and problems of removing the coloured oxidation products from the skin. Encapsulation of the hydrophobic dithranol in surface-active monoglyceride crystals, forming an aqueous dispersion, has resulted in an efficient therapy against psoriasis (Christensen *et al.*, 1989). The crystalline lipid layers covering dithranol are highly effective in protecting it from oxidative degradation (Micanol™, produced by Hydro Pharma).

Plate 13.1. An electron micrograph showing cubosomes, kindly submitted by W. Buchheim and T. Landh. The dispersion was freeze-fractured and freeze-etched. The cubic repetition unit within each cubosome is about 100 Å.

CHAPTER 14

LIPID FUNCTIONALITY IN FOODS

Most foods are colloidal dispersions with gel-like flow properties. The liquid or gel continuum consists of oil or water, and particles or macromolecules in the continuous phase contribute to formation of a gel network. When these two liquid phases occur together, they form oil-in-water or water-in-oil-emulsions. Fat crystals form gel networks in the oil phase, in a similar way as proteins or carbohydrates form gels in the aqueous phase. When gas cells occur in foods, they are dispersed in much the same manner as oil in an aqueous continuous phase.

In food production, a biological tissue is usually first disrupted and then the constituents are separated and combined with other components. In all processing steps, new interfaces are formed, and control of interface formation and structures is really the most critical technology in food production, and a major factor behind food quality. When an interface is formed in complex systems like foods, there is always a competition for this interface in order to reduce the surface free energy. The component which is most surface-active will win this competition, and this is usually a polar lipid. Sometimes, proteins can form the interfaces even in the presence of polar lipids, however. This is due to the irreversible character of adsorption and re-conformation of protein molecules.

From these general aspects of food processing, it is obvious that the functional properties of lipids are a significant field of knowledge in food technology. The physical chemistry of food components is furthermore of interest in nutrition; the bioavailability of certain nutrients depends on the colloidal structure, which in turn is influenced by the lipids present.

A. STARCH MODIFICATION BY INTERACTION WITH POLAR LIPIDS

An important application of polar lipids, "food emulsifiers", in the food industry is based on complexing reactions with starch components in order to modify gel properties. The complex formation with amylose and amylopectin respectively will first be described, and we will then discuss the effect on starch gel formation and gel stability. An extensive review of this field was recently published (Eliasson and Larsson, 1993).

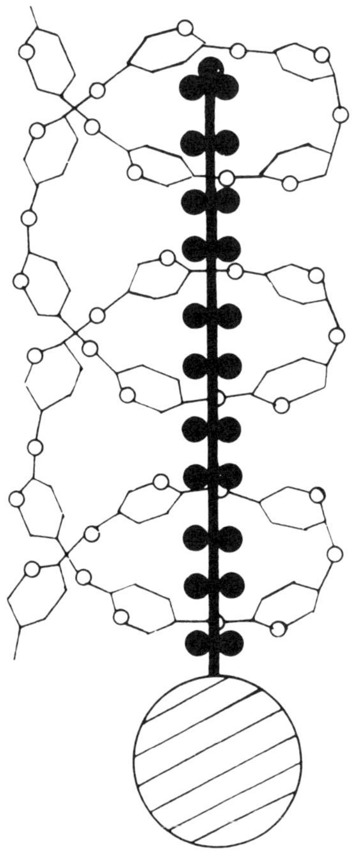

Fig. 14.1. Amylose-monoglyceride complex illustrated by the amylose helix backbone and the polar head and extended hydrocarbon zigzag chain of the monoglyceride, after Carlsson *et al.* (1979).

1. The Amylose-Lipid Inclusion Complex

When amylose in solution is exposed to lipids containing one hydrocarbon chain, an inclusion complex is formed as shown in Figure 14.1. The mechanism behind formation of the complex is the amphiphilic character of amylose; the hydrocarbon chain is hidden in the hydrophobic inside of the amylose helix. This is in fact the only structure of a starch component, that is completely known, as the electron density map of the corresponding complex with iodine has been determined. The helical amylose conformation in these complexes is termed V-amylose. Iodine is also hydrophobic, and the complex contains linear regions of I_3^- or I_5^- (the blue colour of this complex is used to identify starch). The helical amylose chain contains six glucose residues per turn, and from molecular models it is obvious that the interior of the helix is quite hydrophobic, whereas the outside forms a tight hydrogen bond system along the helical tube. The inner space is just big enough to accommodate a hydrocarbon chain, and according to a Raman spectroscopy study, the chain forms the extended zigzag conformation similar to that in crystals (Carlsson *et al.*, 1979).

Amylose in solution forms a random coil conformation, with short sequences of a few helical turns (V-amylose) formed dynamically. This tendency to form a screw is due to the $\alpha(1,4)$-glucoside linkages. Cellulose, with $\beta(1,4)$-bonds, adopts a linear conformation. Polar mono-acyl lipids added to an amylose solution form an insoluble inclusion complex. The amylose-lipid complex precipitating from the solution shows the V-amylose X-ray pattern, corresponding to a parallel packing of the helical units shown in Figure 14.1.

In native cereal starches, lipids dominated by lysophosphatidylcholine are present in amounts ranging from about 1 to 5% (w/w). It has not been possible to record the V-amylose X-ray pattern from this native complex, indicating that it is amorphous (like the arrangement in starch granules of amylose). Depending upon the temperature and the kinetics of formation, it has been possible to differentiate between the formation of an amorphous complex and the truly crystalline V-amylose complex (Biliaderis and Galloway, 1989). The crystalline order is favoured by a high temperature of complexing.

At the present time, the lipids used to modify starch in foods form an insoluble inclusion complex with amylose, and monoglycerides are the most commonly used lipid. In the future though, it seems likely that water-soluble polar lipids will be introduced, for example lysophospholipids. The high solubility of these monomers will change the kinetics of the complexing reaction drastically, and will also affect the physical properties of the complex. There can hardly be any objections to using such lipids from a toxicological point of view, as lysophospholipids are present naturally in starch.

2. *Amylopectin-Lipid Complex Formation*

For a long time, the interaction of lipid with starch was related only to the amylose component, but recently it has been demonstrated that amylopectin can also form a lipid complex (Gudmundsson, 1992). The effect of this reaction is not as obvious as in amylose, and the relative proportion of the molecules that can form a complex is small compared to the situation in amylose.

To understand the possibility of amylopectin complex formation, we first have to consider the structure of the amylopectin molecule, as illustrated in Figure 14.2. The cluster conformation of the $\alpha(1,6)$-branches is related to the tendency to form double-helices (also in solution). This tendency combined with obvious steric factors means that only the outer chain tails of a cluster are available for lipid complexing.

When amylopectin in solution is exposed to a polar lipid, we should

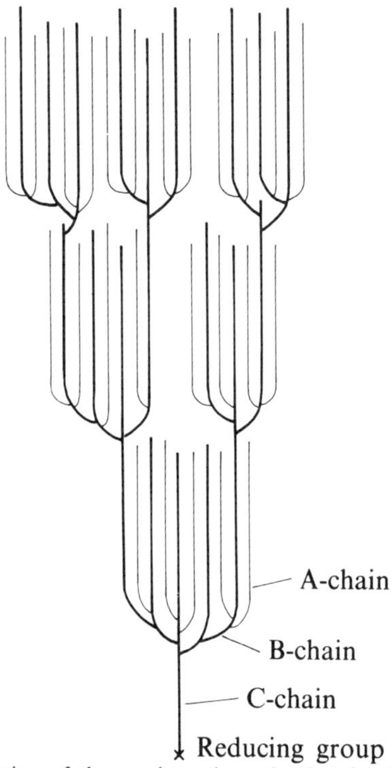

Fig. 14.2. A schematic illustration of the amylopectin molecule, after Eliasson *et al.* (1987). The terms A, B and C show the chain position in relation to the branch structure.

therefore expect that there will be a possibility for formation of helical inclusion complexes by the outer branches. The driving force for this reaction is the reduction in free energy by the V-amylose conformation with its hydrophobic shielding of the lipid acyl chains. If we assume that a certain proportion of such a complex occurs dynamically in solution, for example in the average of five lipid molecules complexed in each cluster, then the crystallization properties of such a cluster will be modified (see next section on starch structure). As in the case of amylose complexation, we must expect that the solubility properties of the complex will depend upon the type of lipid involved. If the lipid is strongly polar, *e.g.* ionic, there should be an increased solubility compared to a complex with a weakly polar lipid. There is not enough space in the helical tube to accommodate a large polar head, which means that the complex will exhibit the properties of the head groups.

An estimate of the proportion of lipid (or surfactant) molecules that is bound to amylopectin in solution has been calculated from surface tension studies (Gudmundsson *et al.*, 1994). The reduction of surface tension of a surfactant molecule is related directly to its concentration up to the critical micellar concentration (CMC), where it becomes constant. A removal of surfactant molecules into an amylopectin complex will be seen

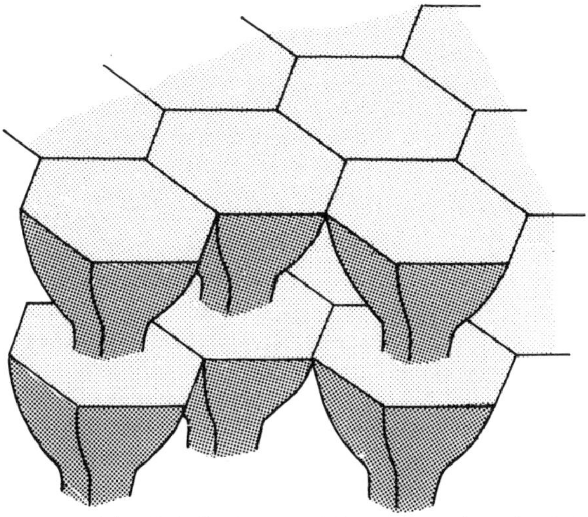

Fig. 14.3 Arrangement of double-helix clusters of the amylopectin molecule in native starch, after Eliasson *et al.* (1987).

directly as a shift in CMC. This study thus demonstrates unambiguously the existence of an amylopectin-lipid complex.

The effects of lipids on starch through complex reaction with amylose and amylopectin involve lipid *monomers*, not the phase of the actual lipid, which is the most common feature of lipid functionality. The phase behaviour of the lipid is still relevant, however, as it will determine the distribution of lipid monomers. The significance of phase behaviour on amylose complexation reactions by different monoglycerides has been clearly demonstrated (Riisom *et al.*, 1984).

3. Starch-Lipid Interactions

In order to understand the possible effects of lipids on starch, we have to start with starch structure. A most remarkable fact is that the crystallinity of the starch granule is due to amylopectin only, whereas amylose occurs only in an amorphous form. The clusters of amylopectin discussed earlier form concentric layers in the starch granule, as shown in Figure 14.3. The molecules are arranged radially in the granule with the branched regions, in clusters, distributed regularly along the main chain. There are about 30 to 40 double helices at the top of the cluster, where it meets adjacent clusters in the semi-crystalline granule. It should alsc be emphasized that the double helices in a cluster of cereal starches are arranged in a close-packed (nearly hexagonal) array, whereas in tuber starches one double-helix out of six is missing and is replaced by an assembly of water molecules. Between the clusters, which form a kind of concentric crystal, amylose in an amorphous state fills the empty space. The amylose-lipid complex also appears to be amorphous (Eliasson and

Larsson, 1993). In most starch varieties, the proportion of amylose to amylopectin is about 1 to 4.

Gelatinization of starch can be obtained by heating or changes in the water structure (induced by urea, for example). It involves the following steps (*cf.* Eliasson and Larsson, 1993):

- leakage of amylose from the amorphous regions between the clusters and replacement of this space by water
- swelling of the granule by additional incorporation of water so that water shells will surround each cluster
- dissociation of the double-helices within clusters by water solubilization

The process is strongly co-operative, and there is some overlap between the steps, because of the variation of curvature of the concentric layers with corresponding differences in lattice energy. If we follow the process in a scanning calorimeter, the gelatinization peak starts at about 60°C. If we perform a corresponding temperature scan with X-ray diffraction recording, it can be seen that the diffraction pattern persists to a much higher temperature. The reason is that the X-ray diffraction pattern is obtained from each cluster; curvature displacement of clusters interrupts a longer periodicity. Even if the clusters dissociate, they will still give rise to X-ray diffraction.

It should be added that there are two more transitions that can occur on heating to higher temperatures. As mentioned above, cereal starches in the native state contain an amylose-lipid complex, and its dissociation can be seen as a transition peak in scanning calorimetry. Furthermore if the water content is too low to allow complete gelatinization, a transition is observed that is assumed to be a kind of melting of ungelatinized crystalline regions. We will not deal more with these two transitions here. A glass transition of somewhere in the interval 0 to 60°C has been demonstrated in starch gels, and the glass description is a significant feature in understanding the physical properties of starch gels (Slade and Levine, 1988).

We are now ready to discuss how the addition of monoacyl lipids can influence starch properties in foods, and we will start with gelatinization. The transport of amylose out from the amorphous regions, when water goes in, can be influenced by the possibility of an amylose-lipid complexing reaction. If we have a monoglyceride present, the complex is not soluble, and the water-filling between the clusters will therefore be hindered. This results is a reduction in the degree of gelatinization that is easy to observe as a reduction in gel volume (provided excess of water is present). Another drastic effect is the influence on gel consistency. The texture thus becomes short, with a reduction in stickiness. This is a property which is utilized often, for example for reduction of stickiness in potato powder or in pasta products.

Alternatively, if we consider an ionic surfactant-type of lipid discussed earlier, we will get quite different effects. Because of the tendency to increase amylose solubility by electrostatic repulsion instead, gelatinization proceeds much further than in the absence of this type of lipid (Kim, 1993). This can be observed as an increase in gel volume, and in fact it is not possible to obtain a gel in equilibrium with external water, even in the presence of enough charged lipid. Another effect comes in here. The complexing reaction with the outer chain tails of the amylopectin molecules of each cluster means that an electrostatic repulsion between clusters will contribute to the water swelling. The opposite situation occurs when we have a less polar lipid like a monoglyceride present. The insoluble character of the V-amylose segment, which is formed, will reduce swelling of the cluster structure and its dissociation.

Industrial applications of charged lipids in the modification of starches have yet to be developed, as mentioned before. Once such lipids become available, we can be sure that applications of this kind will be found. Enzymatic synthesis of food emulsifiers, discussed in the next chapter, will make such strongly polar lipids available.

Finally, we will consider the effects of the storage properties of starch gels. When gelatinized starch is stored, there is a tendency towards recrystallization of double-helices into clusters, and also of clusters into regions like those in native starch. This can easily be followed both by X-ray diffraction recordings and calorimetric measurements. In most cases, this recrystallization, or so-called starch retrogradation, is not desired, and with the addition of lipids it can be delayed and even eliminated. The mechanism behind this modification is probably the formation of amylopectin-lipid complexes at the outer chain tails of the clusters. As these will have the V-amylose conformation, they will break any crystalline packing of double-helices. Lipids such as monoglycerides should be expected to have the same effect as ionic surfactant-like lipids in this respect.

The most common application of reducing amylopectin recrystallization by the addition of polar lipids is probably the use of monoacyl lipids (like monoglycerides) as antistaling agents in bread. The staling process can easily be followed by examination of a piece of the bread crumb by differential scanning calorimetry. A peak (located where amylopectin "crystals" gelatinize) builds up steadily. It can be detected even during the first day of storage. The peak disappears if the bread is heated up to about 70°C. It is also well known that the bread feels fresh in consistency after the same heating process. By the calorimetry method, it is possible to evaluate the effect of different lipids, and the influence of lipid physical state and other factors on bread staling.

Starch gelatinization in bread-making takes place in a water-poor environment, and the starch is therefore not fully gelatinized. In other foods with a higher water content and fully gelatinized starch, storage can result in water separation as a consequence of the same retrogradation process. Pre-manufactured desserts based on starch are an example of this type of product. To increase shelf-life in practice, chemically modified starch is often used. A better alternative with regard to food safety is stabilization of a starch gel by the use of a natural lipid. In this application, the ionic lipids discussed earlier are probably most effective.

B. MILK-BASED LIPID PRODUCTS AND THEIR ANALOGUES

Traditionally, the dairy industry has been the most advanced food industry with regard to research and use of new technology. Milk is certainly the best known emulsion. A very good source of information on the chemistry and physics of milk and milk products is the book by Walstra and Jenners (1984). Before going into details on different types of products, we will state some general facts about milk.

1. Milk

Two different approaches are needed in this context. First we have to regard milk as a biological system, where many biochemical changes occur for some time after secretion from the mammary gland. Lipase activity modifying the lipids and proteolytic enzymes changing the structure of the casein fractions are responsible for the most important biochemical processes, with direct implications on sensory properties. The other approach is to consider milk as a colloidal system. All products from milk are obtained by changes at the colloidal level, as will be seen later.

The milk fat globules, which amount to about 4% (w/w) in cow's milk, have a broad size range with diameters of 1-10 μm predominantly. The milk fat globule membrane is illustrated schematically in Figure 14.4.

When the fat globule is secreted from the mammary gland, it will be enveloped by the plasma membrane of the gland. This membrane has an important role in stabilizing the milk fat globule. A protein-rich layer, with xanthine-oxidase as a major constituent, forms an organized aqueous zone between the polar surface of the milk fat globule and that of the outer membrane. This structure is probably a periodically curved membrane, as discussed in Chapter 9. The milk triglycerides consist of "ordinary" fatty acids provided via the blood, with palmitic and oleic acids being most abundant, and short-chain acids synthesized in the mammary gland. The short-chain fatty acids are located on position *sn*-3 of the glycerol group, and triple-chain-length structures are formed in the corresponding crystal. Crystallization of milk fat is extremely complicated, as the fat consists of so many triglyceride species (Walstra and Jenners, 1984). Even after long times of storage, all three crystal forms, α, β' and β, can coexist.

When milk is homogenized, usually by the action of a pressure gradient

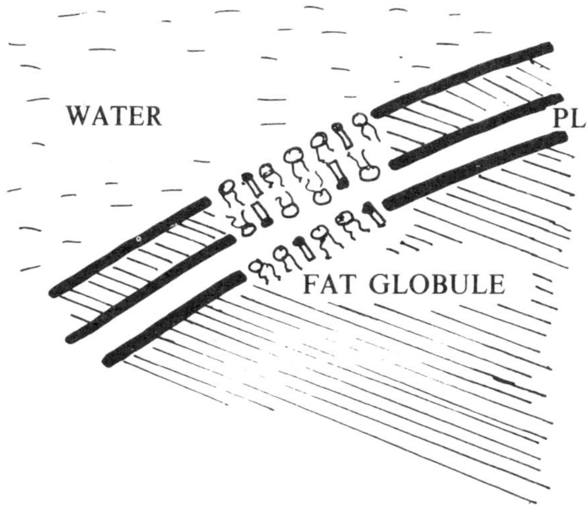

Fig. 14.4. Illustration of the milk fat globule membrane. The protein-rich water layer separating the outside membrane is denoted PL. The phospholipid molecules (with polar heads shown by open circles) and cholesterol molecules (with filled circles illustrating polar heads) are indicated.

in a valve homogenizer, the particles become smaller and more monodisperse. An ordinary homogenization process will give fat globules with about a tenfold increase in surface area. The outer membrane is not sufficient, and caseins will act as emulsifiers. The casein fractions are strongly amphiphilic, especially β-caseins (which even show a critical micellar concentration like surfactants). The stabilization of emulsions by caseins through steric interaction was discussed in Chapter 8.

2. *Whipping of Cream and Production of Ice-cream and Butter*

(*i*). *Whipped cream.* Cream is separated from milk by centrifugation, and a typical cream for food use contains about 30% (w/w) of fat. Whipping of cream cannot give a stabilization of air cells unless fat crystals are present. A critical concentration of fat globules is also needed, as they are directly involved in stabilization of the surfaces of air cells. A two-dimensional matrix of aggregated fat globules can be seen at the gas/liquid interface in whipped cream. The incorporation of air cells during whipping gives a driving force for molecules on the surface of the fat globules to go to the air/water interface. This will result in exposure of triglyceride molecules towards water unless the globules adsorb at the air/water interface. An illustration of the proposed surface structure during whipping is given in Figure 14.5.

We can also regard the process in the following way. When an air cell is introduced into cream, numerous fat globules will attach to its surface. The polar lipids (and also caseins if they are there) covering the globule surface will then spread as a monolayer, and the "naked" part of the triglyceride core will just stay at the interface. With increasing area of the

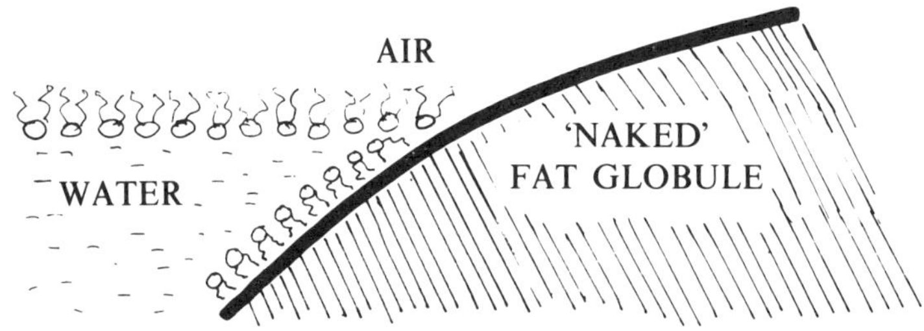

Fig. 14.5. Initial stage of whipping with a fat globule adsorbed at the air/water interface.

interface during whipping, the fat globules will become uncovered to a greater extent. As the globule surface exposes triglycerides, the globules will start to aggregate at the interface. If the core consisted of liquid triglycerides, the globules would immediately fuse. Coalescence is avoided by the presence of fat crystals, which preserve the spherical shape of the globules.

(*ii*). *Butter.* If the whipping is continued after the globules are all aggregated at the gas/liquid interface, the new air cells will remove more of the polar coat of the globules and they will start to aggregate in three dimensions. The result of this is a phase inversion (o/w to w/o) with aggregated fat globules encapsulating some water. This is the mechanism behind formation of butter. The water-in-oil emulsion formed in this way contains about 20% (w/w) of water, and it coexists with milk serum.

The crystal morphology in butter is influenced by the presence of some globule membrane material, and also by the complexity of the triglyceride mixtures. Even the β-crystals, which in other fats grow too large to melt in the mouth, will therefore be sufficiently small.

The formation of suitable crystal sizes and crystal networks is achieved by tempering. More than 50 years ago, a tempering process was developed ("The Alnarp process"), which is now a standard procedure. This was long before anything was known about the crystal forms and their structures. In this tempering process, the temperature of the cream is first lowered to 8°C to obtain an optimal proportion of crystal nuclei. It is then raised to about 19°C for growth of high-melting crystals at the expense of low-melting ones. Additional growth and network formation is then allowed to take place at 12°C, and churning is then started, usually at 10°C. The network of crystals, which is a precondition for formation of a plastic fat, was discussed in Chapter 2. The special features of network formation relevant to butter has been examined by Walstra (*cf.* Walstra and Jenners, 1984).

An alternative to churning has recently been introduced industrially by Alfa Laval (now Tetra Laval). It involves concentration of cream by centrifugation to the composition corresponding to that of butter. At this high concentration, the fat globules attain a polyhedral shape for packing reasons. Because of the strong fat globule membranes, it is still an oil-in-water emulsion. By a minor mechanical disturbance, phase inversion into a water-in-oil emulsion, butter, takes place. An important part of this process is to control the mechanical damage to the fat globule membranes in order to create possibilities for a crystal network formation similar to that in butter prepared conventionally. The obvious advantage of this new technique is that no by-product (buttermilk) is obtained, and the composition of this butter can be pre-determined exactly by the centrifugation procedure. The legal regulations on butter composition are a complication in conventional butter production, as the phase inversion step cannot easily be controlled.

At present, there is an appreciable increase in the number of spreads on the consumer market, based on a mixture of butter fat and vegetable oil. They taste like butter and can be spread at refrigerator temperature. The corresponding process has been called "Alfablend". A typical spread of this type is prepared from a mixture of 75% milk fat and 25% soybean oil or rapeseed oil, with the final product having a ratio of fat to milk serum of 80:20. There are three steps in the Alfablend process. First, the cream is concentrated and stored for "pre-crystallization" at room temperature for several hours. The vegetable oil and salt are then added, and the mixture is fed into a scraped-surface heat exchanger. After phase inversion, the product is worked in pin rotors to obtain the desired consistency.

Finally, it must be mentioned that fat crystallization within fat globules has been found to be different from that in "free" butter fat (Söderberg *et al.*, 1989). This is probably an effect of order induced by the polar lipid surface of the globule acting as sites for nucleation and crystallization promoters, as discussed in Chapter 2.

(*iii*). *Ice-cream.* Ice-cream can be prepared from ordinary cream if about 10% (w/w) of sugar and flavours like vanilla are dissolved in the cream before whipping. The whipping (or some other agitation process) has to be performed under cooling. The crucial step in ice-cream preparation is that the water phase should go over to ice just before phase inversion towards butter takes place. The whole structure with emulsion-stabilized gas cells is thus immobilized because of the presence of ice crystals. The sensory characteristic of a good ice-cream, apart from the cooling sensation on melting of the crystals, is the mouth-feel of the fat phase during its melting and coalescence. The exposure of flavours in the fat and water phases during this phase transformation process gives the final sensory perception.

Ice-cream should contain air cells to about half of its volume, and the ice crystals must be below about 30 μm in diameter, not to give a "sandy" mouth-feel. The structure of ice-cream, its physical properties and aspects of its lipid functionality have been reviewed by Berger (1989). A general feature is the two-dimensional arrangement of fat globules at the surface of the air cells, as in whipped cream. A hydrocolloid, such as a cellulose derivative, is often added to the water phase to reduce growth of ice crystals. Even at a temperature as low as -20°C, all the water is not frozen, and crystal growth can take place.

Today ice-cream is often prepared by using fats and oils of vegetable origin (the name permitted for such products varies between countries with different legislation). When vegetable oils are used, an artificial cream is first prepared, as discussed in the next section. Fat crystals must be present in the emulsion produced, to allow incorporation of air cells. A mixture of a vegetable oil, like soybean oil, and a solid fat, such as a hydrogenated oil or coconut fat, is emulsified by milk proteins (skim milk powder for example). During cooling, the same changes as described above must then take place.

The efficiency of caseins as an emulsifier is a complication in the production of ice-cream. The emulsion should undergo coalescence during melting of fat crystals in the mouth, but too effective surface protection on the fat globules will not allow this. It has been found that if monoglycerides are present in the oil, they will displace the protein surface film to some extent and make coalescence possible (the basic mechanism is described in Chapter 8). By varying its concentration, the added polar lipid can modulate the emulsion stability to a desired value.

3. *Drying of O/W-Emulsions*

Milk and other emulsions are sometimes stored as dry powders. A serious problem in the production process is the tendency of the surface membrane of the oil droplets to rupture during drying, when the external water is exchanged for air. Blending of the emulsion with polar lipids and with carbohydrates has proved to be the most effective way of stabilizing the surface membrane of the emulsion droplets against drying damage. The added polar lipids can coat the polar surface of the fat globules and give them a hydrophobic surface towards air. On reconstitution of the emulsion by exposing the powder to water, the added polar lipids will flip-flop and form liposomes. This shows which type of polar lipids we should use, and PC is the most commonly used lipid today. Carbohydrate also tends to accumulate on the surface of the fat globules during drying. A polar outer environment is created in this way. The polar surface is therefore stabilized, and the emulsion droplets will expose the same surface towards air as they did in the wet stage towards water. Examples of carbohydrates which are used for this purpose are disaccharides and maltodextrins (Whitburn and Dunne, 1991). It is interesting to note that the same mechanism is utilized by cell membranes to protect the native structure *in vivo*. Certain microorganisms are known to produce large amounts of trehalose when they are dehydrated.

Finally, the danger of lipid oxidation occurring during drying must be pointed out. We will return to this problem in connection with nutritional considerations below.

4. Hydrogenation of Oils - Hardened Fats

The melting point of a fat is closely related to the average chain-length of the fatty acids and their degree of unsaturation. If we compare the melting point of tristearin with that of stearic acid, we will find a remarkable similarity (about 70°C). With one *cis*-double bond, the melting points of both drop to below room temperature. The corresponding lipids with *trans*-double bonds have melting points closer to the saturated compounds than to those with *cis*-double bonds. From these general features of melting properties, it is obvious that hydrogenation, which reduces the number of double bonds and changes *cis*- to *trans*-double bonds, will increase the melting point.

Hydrogenation of vegetable oil is a necessary step in order to produce plastic fats for foods, *e.g.* margarine. The process is carried out until a desired melting point is reached; a commonly used hydrogenated ("hardened") rapeseed oil has a melting point of 41°C. The final plastic fat will attain a melting point below that temperature, because of the eutectic effect of the added oil.

The hydrogenation reaction, like all reactions, proceeds in both directions, and therefore the equilibrium mixture of the *trans*-isomers in relation to *cis*-double bonds is obtained. There has been concern about physiologically harmful effects of *trans*-fatty acids recently. In this connection, it should be mentioned that heating of oils during other refining processes can produce *trans*-isomers. The deodorization process, for example, can produce various *trans*-fatty acid isomers, which are different from those obtained by hydrogenation. Oleic acid will give some elaidic acid after partial hydrogenation, and later heating can result in movement of this *trans*-double bond along the fatty acyl chain.

If we want to produce a plastic fat that does not contain *trans* fatty acids, one possibility is to perform total hydrogenation of the oil. The melting point can then be reduced by interesterification with a fat containing *cis*-double bonds. This can either be done chemically or enzymatically. Assume that we add equal amount of triolein to tristearin. Chemical interesterification will then result in a random mixture of all isomers (OSS, SOS, OOS and OSO). When interesterification is achieved enzymatically, the fatty acid exchange takes place with specificity for positions 1 and 3. In the last chapter, we will consider a margarine produced in this way.

5. Artificial Milk and Cream. Margarine and Low-Fat Spread Products

We have discussed artificial cream above in connection with ice-cream. Reasons for preparation of milk or cream based on vegetable fats and oils are a lack of milk, for example in tropical regions, and the lower costs of raw-materials. Unless whippability of the cream is desired, it is not necessary to have fat crystals present in the artificial milk or cream. A straight-forward emulsification, as described in Chapter 8, can therefore be applied.

An interesting application of polar lipid phase behaviour is the use of an L3-phase to improve the whippability of artificial cream. Monoglycerides of unsaturated fatty acids, for example those prepared from sunflower oil, have been found to improve the whipping properties. By examination of the phase diagrams of the corresponding oil/water/monoglyceride system (see Chapter 3) and the effect of milk proteins, it is possible to understand what happens at the interfaces during whipping.

The effect of caseins on the cubic monoglyceride-water phase has been described (Buchheim and Larsson, 1987). Caseinate is able to disperse the cubic phase in excess of water, but the particles are not cubosomes, as evident from their EM-appearance. A bicontinuous structure like that of cubic phases was observed, but with a much longer repetition period. Our interpretation was of a somewhat disordered gyroid-type of structure. Since then the L3-phase has been analysed. It seems very probable now that the sponge-like structure we observed is actually an L3-phase, *e.g.* a disordered cubic arrangement of an infinite monoglyceride-caseinate bilayer. This L3-phase needs additional caseinate for its dispersion in water; these molecules will probably be located at the bilayer fractures. An artificial cream is prepared by emulsification of unsaturated monoglycerides in the oil phase (concentrations of several percent are used) and caseinate (a few percent) in the water phase. This emulsion will contain L3-phase regions. When air cells are whipped into this emulsion, an amphiphile monolayer will form at the air/water interface in order to reduce the surface free energy. The high monomer concentration of caseinate in the aqueous phase compared to that of the monoglyceride (about 10^{-6}) means that caseinate will dominate this new interface. With removal of casein, the dispersed L3-phase will start to expose hydrophobic regions, which then will tend to associate with fat globules. The result will resemble that described in Section 2 (*iii*) above.

Various other food emulsifiers are used in the production of artificial milk and cream. Stearoyl-lactyl-lactates, for example, have proven to be good emulsifiers, probably because of the electrostatic repulsion they induce between the droplets. Palm oil is increasingly used as a plastic fat in foods, and a problem has been the formation of so-called granular crystals. From model studies, it has been shown that the granular crystals are formed by the POP-component in a triple-chain-length β-form (Walstra *et al.*, 1992).

6. *Margarines*

The preparation of margarine was initiated by Napoleon III as a substitute for butter. Today margarine is superior to butter with respect to its storage and spreading properties, and perhaps also with regard to nutritional aspects if low-caloric products are included.

An ordinary margarine is prepared simply by feeding a mixture of an oil phase and a water phase into a scraped surface heat exchanger. A feature article on margarine technology has been published recently (Rajah, 1992). It contains a full description of the engineering aspects of the production of margarines and other spreads. Crystallization of parts of the fat is the main factor responsible for stabilization of the water droplets in the fat phase. Usually, the water phase is a soured skim milk, and an emulsifier is added to the oil phase. One important function of the emulsifier is to give the water droplets a uniform size that is sufficiently small. The sensory properties, for example the taste of added salt, depend on the size of the water droplets. The most common emulsifiers used are different oil seed "lecithins", *i.e.* the complex phospholipid mixture obtained when the extracted oil is exposed to water or acid (in the "degumming" of the oil) as the first step in refinement of the oil. Soybean oil lecithins are the most common ones, and they are generally added in fractions of a percent to the oil.

Margarines have traditionally been prepared with about 20% water to correspond to the composition of butter. During the last two decades, margarines with a lower fat content have steadily increased their market share, however. In order to prepare margarines with 60% and 40% (w/w) of water, it is necessary to add other emulsifiers. A margarine with 20% of water will exhibit water droplets in sizes of around 5 μm in diameter, whereas an increase in water content to 60% gives water droplets about five times larger. Distilled monoglycerides are the component used most frequently to obtain the desired size distribution of the water droplets. With a high water content, it is necessary to increase the viscosity of the water phase in order to further stabilize the emulsion. Milk proteins and gelatine are used. The main problem with low-fat spreads, with a fat content of 20% or even less, is to obtain a uniform water-in-oil emulsion. Many products of this type on the market appear to be inhomogeneous or double-emulsions, and the presence of open water regions opens the possibility for spoilage by microorganisms.

The different types of margarines discussed above require a fat crystal network in the oil phase, and experience has shown that only the β'-form has an acceptable morphology and size distribution. The presence of β-crystal results in a "sandy" mouth-feel. In order to inhibit the $\beta' \rightarrow \beta$ transition during the shelf-life of the margarine, the hydrogenated fat must have certain fatty acid characteristics. If about 80% or more of the fatty acids have the same chain-length, the hydrogenated fat will be strongly β-tending. The problem of changed polymorphic behaviour of hydrogenated rapeseed oil with a low content of erucic acid (LOBRA) was discovered more than 20 years ago by Riiner (1971). Since then, the desired heterogeneity in chain-length distribution has been achieved by plant breeding.

There is also a possibility of adding lipids which by co-crystallization will reduce the kinetics of the $\beta' \rightarrow \beta$ transition. In the case of hardened rapeseed oil, the use of diglycerides was found to be effective (Hernqvist *et al.*, 1981). Surfactants have been found to be very efficacious in delaying triglyceride polymorphic transitions (Aronshime *et al.*, 1990). The effects of sorbitan-esters on tripalmitin polymorphism were recently reported on the basis of phase diagram studies (Mohamed and Larsson, 1992). Furthermore the binary system sorbitan-tristearate/tristearine is under detailed investigation, demonstrating a high molecular solubility of the sorbitan ester in the triglyceride crystal lattice (Elisabettini, personal communication). The molecular solubility of the sorbitan ester is probably the reason behind the delay in the $\beta' \rightarrow \beta$ transition. It should also be mentioned that sorbitan-monopalmitate in fact promotes the phase transitions $\alpha \rightarrow \beta' \rightarrow \beta$ in tripalmitin. This is related to the apparent formation of a sorbitan-monopalmitate/tripalmitate 1:1 molecular compound.

We should expect retardation effects on phase transitions, and thus stabilization of the β'-form under the following conditions:

- co-crystallization into the lattice is possible for a substance with a strong tendency to have its chains packed according to the orthorhombic chain packing subcell. This is probably the effect obtained by diglycerides.
- co-crystallization of a lipid which makes the methyl end group plane more disordered. This will reduce the gain in close-packing by phase-transitions into the more stable polymorphic form.

The best alternative is certainly to "tailor" the high-melting triglycerides of a margarine so that they will be sufficiently β'-stable. The main factor is the degree of heterogeneity of fatty acid chain-lengths. By experience, we know that if as much as 80% of the fatty acids have C_{18} chains, then hydrogenation will produce a fat which is not sufficiently β'-stable.

The association of crystals into a network in the oil phase is discussed in Chapter 2. The taste of margarine can of course be varied, yet it appears as if only butter aroma substances are used. In our department, we have prepared margarines in which about 20-30% of the oil phase consists of virgin olive oil. Such spreads tasting of olive oil have been well accepted in sensory evaluations. There are numerous alternatives for flavouring margarine, that do not seem to have been explored.

Finally, it can be mentioned that special margarines have been developed for certain applications. Baking margarines should contain fully hydrogenated fats for use in continuous bread production lines. (Fats in baking will be discussed in the next section). When a margarine is used for frying, it is important that the water droplets remain dispersed even after melting of the fat crystals to avoid spattering. The use of emulsifiers as effective anti-spattering agents is therefore needed.

The current market for butter, margarines and spreads and legal aspects of their classification have recently been reviewed (Nichols, 1993).

C. COCOA BUTTER AND CHOCOLATE

The cucumber-like fruits from the cacao tree were first used in Central America and Mexico. Because of their high value, they were used for payment, as reflected by the original Latin name *Amygdala pecuniaria*. The tree was later named *Theobroma cacao* by Carl von Linne.

Cocoa is the name used for fermented beans. After roasting, the cocoa butter is separated from the cocoa powder. The powder is predominantly proteins, but contains 10-20% (w/w) of cocoa butter nevertheless.

Chocolate consists of cocoa butter as the continuous phase with dispersed particles of sugar and cocoa powder (non-fat cocoa constituents giving the characteristic flavour). Phospholipid mixtures are used as dispersion agents. As cocoa butter is expensive and sometimes not available in sufficient amounts, cocoa butter substitutes have been developed. Whether such replacers can be used or not in chocolate varies between countries. In countries, where they are not permitted, they can still be used in chocolate-like coatings. For the ordinary consumer, this is the same product as chocolate.

The most characteristic property of chocolate is its melting behaviour. It is solid up to 25°C and fully melted at 35°C. Furthermore, the brittle consistency of a piece of chocolate and its glossy appearance are due to unique properties of the crystal matrix of cocoa fat. It is therefore natural to discuss this component first.

Cocoa butter or fat consists of a very homogeneous triglyceride mixture, with oleic (O) acid in the 2-position and stearic (S) or palmitic (P) acid in the 1- and 3-positions. Thus, cocoa butter contains mainly SOS, POP and SOP, and all these types of molecules appear to be able to enter the same crystal lattice. Still the polymorphism is quite complicated, with the occurrence of six crystal forms. The phase transitions were described by Wille and Lutton (1966), and the crystal forms were classified I to VI. It is now appropriate to classify them by the same nomenclature system as defined in Chapter 2, and then they are identified by the following short-spacing data (given in Å with the relative intensity as very strong (vs), strong (s), medium (m), weak (w), and very weak (vw)).

sub-α	3.87(m), 4.17(s)
α	4.20(vs)
β_2'	3.87(vw), 4.20(vs)
β_1'	3.75(m), 3.88(w), 4.13(s), 4.32(s)
β_2	3.65(s), 3.73(s), 3.87(w), 3.98(s), 4.22(w), 4.58(vs), 5.13(w), 5.38(m)
β_1	3.67(s), 3.84(m), 4.01(w), 4.21(vw), 4.53(vs), 5.09(vw), 5.37(m)

The most stable form is thus the β_1-form.

When cocoa butter is crystallized to give chocolate, it is of great importance that the most stable form (β_1) is obtained. The melting characteristics will otherwise not be correct, and recrystallization is likely to occur during storage. Such recrystallization will result in a greyish rough surface, 'blooming", and such a product is not acceptable.

Sato and co-workers introduced a method of promoting crystallization towards the most stable form (*cf.* Garti and Sato, 1988). This is particularly useful in the case of cocoa butter substitutes, as will be discussed below. They added the homologous 2-oleo-1,3-dibehenin. It has a higher melting point than its homologues in cocoa butter, and it will therefore provide crystal nuclei towards the β_1-form.

As discussed in connection with butter, it is necessary to apply a tempering step in order to first produce nuclei of the most stable form and then a controlled crystal growth. The melt is cooled initially to produce nuclei. Then the temperature is increased so that nuclei of forms other than the β_1-form melt. The temperature is then lowered to obtain final crystallization after moulding.

In the production of dark chocolate or milk chocolate, the components are mixed, and by means of a refining process the particle size is reduced to get a smooth paste. The size of the cocoa powder particles must be less than 30-50 μm, and the sugar crystals must be less than about 35 μm. The fluid chocolate paste is then "conched", a process named from the machines used. At a temperature of about 60°C, the fluid mass is exposed to air for many hours in a kind of stirring and mixing process. The purpose is to produce the desired flavour by getting rid of certain off-flavours. A typical composition of milk chocolate is about 30% milk powder, 20% cocoa mass (with roughly equal amounts of cocoa butter and cocoa powder) and 50% sugar. The moisture content is about 1% and the lecithin level roughly 0.5% (Minifie, 1970).

The function of the phospholipids is to disperse the hydrophilic solids and to provide the desired viscosity for conching and the moulding process. The phospholipids will cover the sugar particles in the same way as discussed in connection with emulsions. By exposing the polar head towards the sugar crystal and the hydrocarbon chain tails outwards, a coating of a phospholipid monolayer will confer hydrophobic properties on the sugar crystal, and this is a prerequisite for dispersion.

Vegetable "lecithins" used in chocolate production are usually prepared from soybean oil, and PC is the most abundant component. Apart from forming a monolayer on hydrophilic particles, its role is not fully understood. It also influences the fat crystallization process and contributes to bloom resistance. Furthermore it influences the gloss of the surface. The level of lecithins added varies in the range 0.2 to 0.6% relative to the total weight of chocolate (Minifie, 1970).

The sugar is milled to a powder with sizes in the range 25 to 35 μm, and the milling also produces a certain proportion of amorphous sugar, which is important for aroma entrapment.

Milk fat is often added in chocolate production. It will lower the melting point somewhat, because of a eutectic effect. However, it will not disturb the polymorphic behaviour of cocoa butter, as there is almost no co-crystallization.

Cocoa butter substitutes must simulate the very specific melting properties of cocoa butter. Cocoa butter alternatives (CBA) can be divided into:

- cocoa butter equivalents (CBE), which contain very similar triglycerides, and therefore they should be tempered in the same way as cocoa butter
- cocoa butter replacers (CBR) and cocoa butter substitutes (CBS), which consist of different triglycerides, and therefore do not require to be tempered like cocoa butter.

A good review of vegetable fats as replacement for cocoa butter was published recently by Rossell (1992). He classified replacers in four groups:

- lauric fats
- hydrogenated fat fractions
- fats with oleic acid in the 2-position ("cocoa butter equivalents")
- novel fats with reduced calorie contributions

Lauric fats (triglycerides rich in lauric acid) are mainly palm kernel oil and coconut oil. As the shorter chains present in coconut oil gives disturbing aroma effects, palm kernel oil is the most important source. It has too low a melting point, however, and must be fractionated or hardened. Both these fats work well in coatings, but they should not be mixed with cocoa butter, as they crystallize in a β'-form. They are therefore used when a total cocoa butter replacement is desired. Any lipase-activity in the actual product will result in a soap-taste, which limits the application of lauric fats.

Hydrogenated fats are the most common substitutes for cocoa butter used at present; hydrogenated rapeseed oil is one example. Such fats give β'-forms of relatively high stability. They can still be mixed with cocoa butter to a certain extent, especially if crystallization is modified by adding other lipids. The use of 2-oleo-1,3-dibehenin is a promising alternative for such products, and sorbitan tristearate is also used.

Cocoa butter equivalents are now produced industrially by enzymatic synthesis of triglycerides. The chemistry of this process will be discussed further in the last chapter. A good starting material should be rich in oleic acid in the 2-position, so palm oil, olive oil and new sunflower oil varieties can be used. Stearic acid is added, and because of the specificity of the lipase, the stearic acid will go into the 1- and 3-positions. On an industrial scale, one problem which is not yet solved is the satisfactory removal of diglycerides and fatty acids produced by the reaction.

A low caloric fat called caprenin has been introduced on the US market. It consist of a triglyceride mixture, in which two chains of the molecule are capric (C_{10}) and/or caprylic (C_8), while the third chain is behenic acid (C_{20}). The melting characteristics resemble those of cocoa butter. A most remarkable property is its low caloric content; it has been found to provide only 5 calories per gram, compared to about 9 calories per gram for ordinary fats. It seems likely that the reason is poor micellar solubilization of the behenic acid and monobehenin produced by the intestinal lipolysis.

The EU and US legislation on the use of vegetable fats in chocolate has recently been summarized (Kernon, 1992). Also, it should be mentioned that there has been considerable work on the possibility of replacing vegetable "lecithin" by other polar lipids. Thus, ammonium salts of phosphatidic acid, sucrose esters and polyglyceryl polyricinoleate have been tried, but vegetable lecithins still appear to be superior (Minifie, 1970).

Production of chocolate and the development of cocoa butter substitutes are illustrative examples of food technology with a scientific basis. The present applications have utilized knowledge of the relationships between the chemical structure of the triglyceride molecules and physical properties as determined by molecular organization. We can now describe the structure of a piece of chocolate from the colloidal level down to the molecular level. The crystal structure of the cocoa butter continuum is a triple-chain length structure (see Chapter 2). There is a small amount of triglyceride oil between the fat crystals, which is space-filling like mortar between bricks. The dispersed solid particles are covered by a monolayer coat of phospholipid, and phospholipid molecules in excess are dispersed as liquid-crystalline aggregates with a hydrophobic exterior.

D. VEGETABLE OILS, FRYING OILS AND O/W-EMULSION PRODUCTS

1. Frying Oils

Foods are cooked by heating in water or in oil, and the differences in surface temperature and chemical environment result in quite different characteristics for the food products. A serious problem in the use of oils for frying is the oxidative changes, and the limited life-time of the oil in order to be acceptable from a nutritional and sensory point of view. It is obvious that a high degree of saturation of the hydrocarbon chains will increase this stability. On the other hand, some oil will always penetrate into the product and from a nutritional point of view, highly saturated oils should be avoided.

A common method for quality control in oils is determination of the amount of free fatty acids present. Other tests are related to the physical properties of the oxidation products; one is based on the foaming of the oil and another on its viscosity.

A new interesting aspect of frying oil is diffusion on the surface of the product and the efficiency of heat transfer (Yasukawa and Aigami, 1991). Through the presence of phospholipids, the amount of cooking oil needed could be reduced by a factor of two. In this context, the disadvantage of using PC is the presence of the amino group, which gives a negative aroma effect. This can be eliminated by enzymatic removal of the choline moiety (Kawahara, 1993).

2. *Mayonnaise and Dressing Products*

Mayonnaise is an oil-in-water emulsion, with egg yolk used as emulsifier. Vinegar provides an antimicrobial effect by lowering the pH to about 4. The most remarkable structural feature of this emulsion is the volume fraction of oil. It is usually much higher than that corresponding to close-packing of spheres (according to the US standard, the minimum volume fraction of oil is 65%, and close-packed spheres have a volume fraction of about 74%). Because of this polyhedral distortion of the oil droplets, such an emulsion might be termed a biliquid foam. When the oil droplets are viewed in the polarizing microscope, a birefringent shell corresponding to the L_α-phase can be seen (lipid multilayers). The interfacial properties of the L_α-phase (see Chapter 8) explain the remarkable stability of the emulsion, an obvious example of a phase forming the oil/water interface.

Dressing products are also o/w-emulsions with a liquid oil phase, but the volume fraction of oil is lower (a minimum of 30% according to the US standard). The oil droplets should have a similar size distribution as in mayonnaise, which means a diameter of 1-5 μm. When egg yolk is used as emulsifier, the interface consists of egg yolk lipoproteins forming an L_α-phase. There is a trend at present to develop "light" products with a reduced oil content. To retain the consistency and mouth-feel, the obvious solution is to add hydrocolloids.

The possibility of replacing egg yolk as an emulsifier will be considered here. There are two strong reasons for this. The most important one is the risk of the presence of pathogenic microorganisms (Salmonella). A mild temperature treatment can be applied, which does not destroy the functional properties of the egg yolk, but some uncertainty about the microbial status may remain. A second reason to replace egg yolk is related to the frequent use of dried egg yolk in industrial production. As discussed in Section F below, there is then a risk of oxidation of cholesterol, a component that is clearly dangerous in a food product. Mayonnaise and dressing products usually contain some air, introduced in the emulsification process, and oxidative changes can take place during storage.

If we want to get rid of cholesterol, the theory of emulsion stabilization in Chapter 8 can be applied. One alternative is to use phospholipids, which form an L_α-phase with water. A PC-rich fraction of plant phospholipids can be used, for example. The other alternative is to use caseins, caseinate or milk proteins rich in caseins. Both of these alternatives for emulsion stabilization have the advantage that the final product can be subjected to heat treatment before filling.

The physico-chemical aspects of spoilage of food emulsions, such as mayonnaise, have recently been analysed (Brocklehurst *et al.*, 1993). The preservation of mayonnaise and salad dressings is related to the antimicrobial effect of acids like acetic acid or added preservatives like benzoic acid, which must be in the undissociated form to be effective. A problem is that the undissociated form is lipophilic and therefore tends to dissolve in the oil phase. The partition of acetic acid into the oil is, however, very low. In the case of added preservatives, such as sorbic acid or benzoic acid, the partition into the oil is significant and must be taken into consideration.

E. LIPIDS IN BREAD AND OTHER BAKED PRODUCTS

1. Wheat Flour Components and their Interaction in Water

Wheat flour consists of about 65% (w/w) starch, 12% gluten proteins, 14% water, 2% lipids, and the rest is non-starch carbohydrates, non-gluten proteins and inorganic salts. It is well-known that the unique baking performance of wheat is due to the functional properties of the gluten proteins, the ability to form an elastic dough with a gas-holding capacity. Gluten is defined functionally as the gel obtained after water washing of a wheat flour-water dough under mixing, when water-soluble components and starch are removed. The gel is rubber-like and also contains the wheat lipids (about 8% of the dry-weight). The proteins are extremely heterogeneous, and consist of two main groups: gliadins, which are soluble in ethanol (70%) solution and glutenins, which are the insoluble remainder. The range in molecular weight is 30,000 to 80,000 Daltons in gliadins and 30,000 up to several million Daltons in glutenins. Both groups of proteins are extremely amphiphilic, and they exhibit a phase behaviour which is quite similar to that of polar lipids. Thus, the pure fractions as well as their complex mixtures swell in water into homogeneous protein-water phases with a well-defined limit of swelling. Proteins of the glutenin group form intramolecular disulphide bridges, and the gel strength is directly related to this cross-linking (Eliasson and Larsson, 1993).

The interaction of wheat lipids with water is demonstrated by the ternary phase diagram shown in Figure 14.6. About one third of the wheat lipids is lyso-phospholipids, which form an inclusion complex in

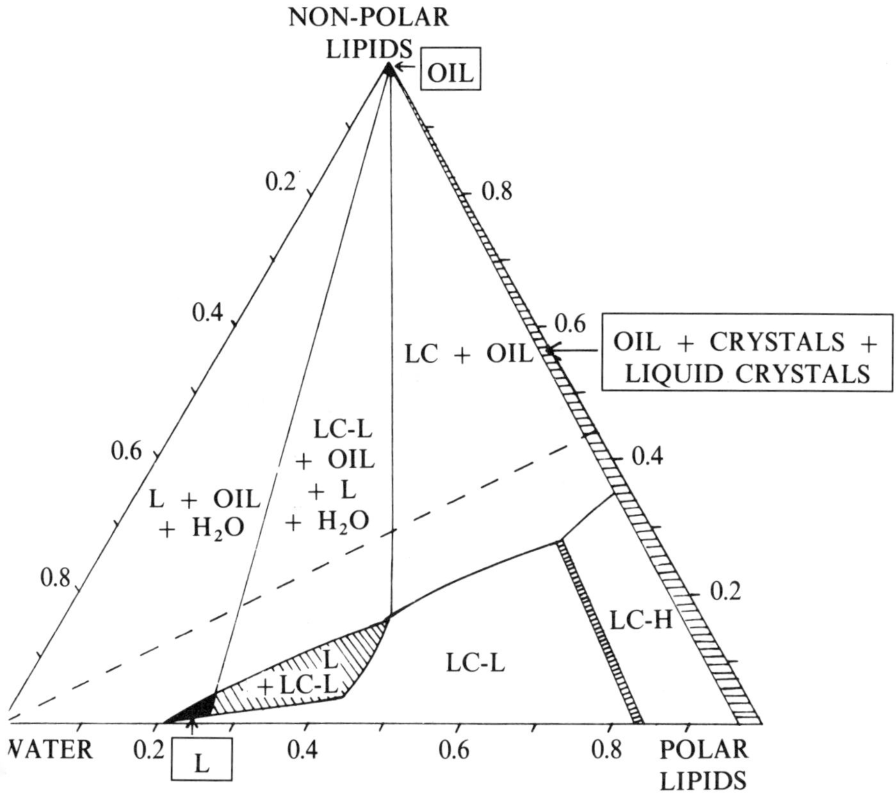

Fig. 14.6. Ternary phase diagram of wheat lipid polar and non-polar components and water (*cf.* Eliasson and Larsson, 1993). The liquid crystalline phases (LC) are the H_{II}-phase (-H) and the L_α-phase (-L), and the liquid phases are oil, water, and probably L3 (L).

starch (discussed in Section A). The rest is a mixture of non-polar lipids, mainly triglyceride oil, and polar lipids, mainly galactolipids and phospholipids. When the phase diagram was determined, we separated the extracted total lipids, in order to find out the effects of polar and non-polar lipids on phase behaviour. As polar and non-polar lipids are distributed in different tissues in the wheat endosperm, the mixing with water when a dough is prepared does not result in an equilibrium mixture; therefore, we must consider the whole phase diagram in order to understand all features of lipid functionality.

The dotted line in the phase diagram corresponds to the composition of the lipids in the flour. Following this line towards the water corner, we can see that the excess of water relevant to dough mixing gives three liquid phases in equilibrium, oil, water and a liquid phase which today should be called L3 (see Chapter 3). There is strong evidence for the requirement of an L_α-phase in the diagram in order to have acceptable baking performance (Eliasson and Larsson, 1993).

Starch behaves as an inert filler in dough mixing; the granules are

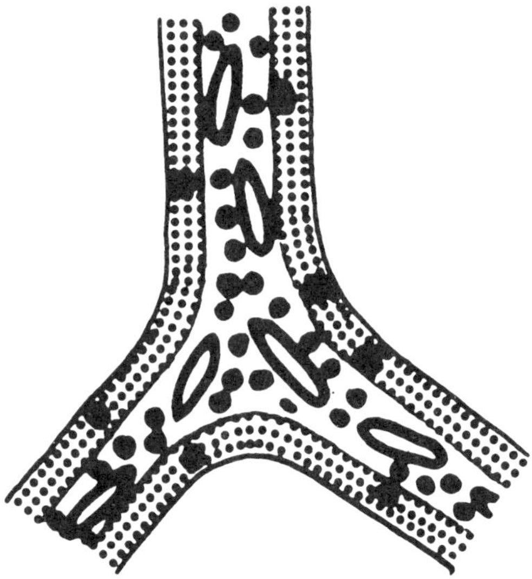

Fig. 14.7. Structure of the foam lamellae in a dough, after Eliasson and Larsson (1993). The gluten gel is shown by the dotted area, and black regions indicate lipid aggregates. Large and small starch granules in the centre region are also indicated.

hydrophilic and tend to be surrounded by water. Fermentation of the dough results in degradation of starch and formation of ethanol and carbon dioxide.

2. *Bread-making*

Ordinary white bread is perhaps our most important food product. It is prepared by mixing wheat flour with water and yeast (typical weight proportions of 100:65:2). There are other optional components, like fat, emulsifiers and sugar, but they will not be considered in this context. The effect of added fat and emulsifiers can be fully understood from the discussion of functionality of the lipids below. The mixing process results in the incorporation of small air cells, which act as nuclei for the carbon dioxide formed during the fermentation. If we cut through a ready fermented dough with a razor blade, it looks like a foam. The structure is indicated in Figure 14.7. The surface monolayer of the foam lamellae consists of the high-molecular-weight (HMW) fraction of the glutenin molecules, and inside this surface monolayer there is a surface zone consisting of the gluten phase (Eliasson and Larsson, 1993). The most important aspect of this new model of the dough is the occurrence of two-phase separation into a gluten phase and a starch-water phase. Liposome-like lipid aggregates also occur in this surface zone formed by the gluten gel. The core of the foam lamellae is formed by the starch-water phase.

When the dough is heated in the oven, the gas cells expand, and the

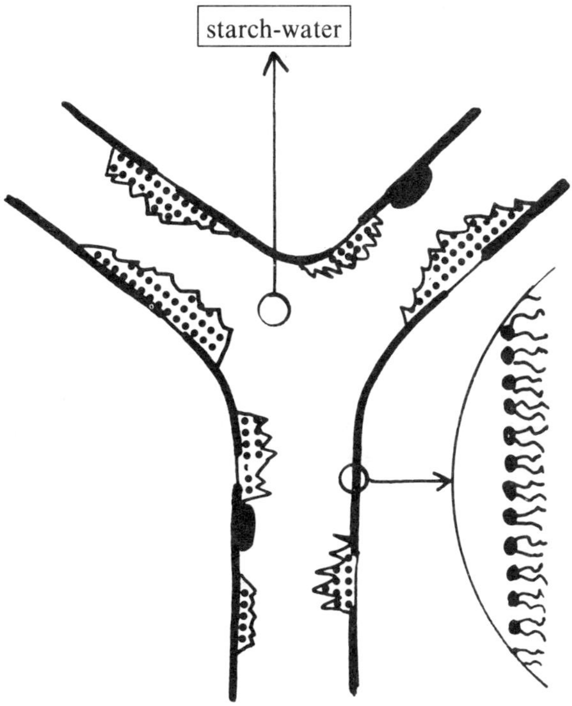

Fig. 14.8. Schematic illustration of the initial structure of the foam lamellae formed during heat expansion of the gas cells in the oven, after Eliasson and Larsson (1993). The gluten gel is indicated by dotted areas.

polar wheat lipids play a crucial role in this process. The structural changes in the foam lamellae, compared to the situation in the dough discussed above, are shown in Figure 14.8. The surface consists mainly of a lipid monolayer, with only patches of HMW-units of glutenin. The gluten gel looses its cohesiveness and elasticity, and the starch gelatinization process leads to drying of the gluten gel. The most prominent change related to these molecular phenomena is the fusion of gas cells, which results in an open pore system in the final bread.

It can be mentioned that the high surface temperature leads to a compact crust structure. At this elevated temperature, chemical changes give the characteristic colour and flavour.

The fixation of the bread structure by starch gelatinization cannot take place unless the polar lipids first have stabilized the gas/gel interface during the rapid expansion of gas cells. The polar lipids must occur in the L_α-conformation in order to be dispersed well enough in the surface zone of the dough. The balance between different lipid species is one critical factor in forming such a phase, especially the balance of MGDG to DGDG (*cf.* Eliasson and Larsson, 1993). When lipids or "emulsifiers" are added to a dough to improve baking performance, they should therefore favour the L_α-phase at the expense of the H_{II} and cubic phases. The additives that by trial-and-error have been found to work best, for example diacetyltartaric acid esters, form an L_α-phase with water. This knowledge is also important in plant breeding; a high proportion of DGDG is as significant as the HMW-glutenins in order to give an acceptable bread.

Fats in the form of baking margarines or shortenings (which do not contain water) are often added to the dough up to a level of 1-2% (w/w). In fact, the presence of fat crystals is required in the continuous process of dough mixing applied in industrial baking. The functional effect of the fat crystals is a strengthening of the foam lamellae, and the various polymorphic forms perform differently in this respect (*cf.* Eliasson and Larsson, 1993). The best effects with regard to bread volume are achieved when the fat crystals do not melt until the structure is fixed by starch gelatinization. This in its turn means that the fat crystals formed after cooling of the bread will not melt at the temperature in the mouth. Their largest size must not exceed about 30 μm, therefore. The fat tends to cover the surface of the pore system after melting (in order to reduce the surface energy). Thus, it has a protective effect against drying of the bread, and it acts as a solvent for lipophilic aroma compounds.

3. Cakes and Cookies (Biscuits)

Cakes are baked from a batter containing predominantly wheat flour, egg, sugar, more or less fat and a gas-forming agent. The physical and structural basis of the cake production process has been well analysed (Shepherd and Yoell, 1976). The heating gives a porous texture like that described in bread above. The mechanism behind the formation of this structure is different, however.

When sugar is mixed into the flour-fat paste, air cells are incorporated. The addition of egg and milk results in solution of the sugar crystals and the flour particles become dispersed in the aqueous phase.

During baking, the first changes are due to the melting of fat crystals in the range 35 to 40°C. The air cells are first stabilized by fat crystals at the oil/air interface. When the fat crystals melt, the air cells move from the oil phase to the water phase (where egg proteins will stabilize the air/water interface). This volume increase of the water phase also contributes to the phase inversion, the formation of an o/w-emulsion.

During further heating, the air cells expand because of formation of carbon dioxide and water vapour. The final development of the cake structure takes place in connection with starch gelatinization and protein denaturation in the range 60 to 70°C, and fusion of the expanding gas cells into a pore system at close to 100°C, where the structure is fixed finally.

Other types of cakes are formed by similar mechanisms, even fat-free cakes (*cf.* Mizukoshi, 1989). It can also be mentioned that lipids can replace egg yolk as emulsifier. Hydrophilic lipid crystal dispersions are especially effective in their ability to stabilize air cells.

Cookies (biscuits) are baked from high-fat batters; characteristic proportions are wheat flour - 100 g, margarine (or butter) - 150 g and sugar - 50 g (with various other ingredients such as cocoa powder and baking powder). When the fat phase is mixed with the flour and sugar, the particles are dispersed in the fat with a tendency to fuse with the water droplets. Air cells are incorporated into the continuous oil phase and are kept immobilized by the fat crystals. During heating in the oven, there is a phase inversion induced by melting of fat crystals, when the air cells move into the water phase, in a similar way to that described in the case of cakes. The surface of the pore system is covered by an oil/fat film, which also functions as a carrier for aroma substances formed by Maillard reactions in the oven.

F. THE RELATIONSHIP BETWEEN LIPID OXIDATION AND MOLECULAR ORGANIZATION

1. General Features

Chemical changes of lipids are of the utmost importance in living tissues, for example through formation of second messengers by phospholipase attack on PI or biosynthesis of prostaglandins from arachidonic acid. Oxidation of fats and other lipids is a most important chemical change *in vitro*. The oxidation state of fats and oils is the most important quality factor in the production and storage of foods. Below, we will consider the oxidation of lipids under conditions *in vitro*, with special regard to the effect of different molecular arrangements.

The significance of molecular organization in the protection of the living cell against lipid oxidation is obvious, if we consider the interplay of tocopherol and ascorbic acid. Lack of tocopherol in cell membranes results in a loss of the barrier properties of the lipid bilayer; erythrocytes start to leak haemoglobin, for example. The tocopherol molecule is certainly arranged with its terminal hydroxyl group in the polar sheet of the lipid bilayer. There is a pronounced deficiency of hydrogen donors for hydrogen bond association in the lipid bilayer of membranes containing mainly lipids such as PC, which has acceptor groups for hydrogen bonds only.

It is well-known that the lipid oxidation in the bilayer is hindered by tocopherol by means of formation of a tocopherol radical. Tocopherol is then regenerated by a reaction with ascorbic acid, located in water outside the bilayer. The ascorbic acid in its turn is regenerated by different redox systems, such as glutathione/glutathione peroxidase. The formation of the tocopherol radical will modify the hydrogen bond association of the polar sheet of the bilayer, and the detailed organization is certainly a significant factor in the reaction with ascorbic acid in the water phase.

From one-electron reduction potentials of the different free radical reactions possible, it has been possible to derive a 'pecking order" of the different species in lipid oxidation, that can occur in a food system (Buettner, 1993). The hydroxyl radical is at the top of the list and can therefore oxidize all the other reduced species. α-Tocopherol and ascorbic acid are at the bottom. In addition, as they are regenerated, they provide a very efficient antioxidant system.

Only free fatty acids are attacked during enzymatic oxidation by lipoxygenases. A common change during the storage of lipid-containing foods is enzymatic hydrolysis by lipases. A lipase requires an oil/water interface in order to have reasonable activity, so the colloidal structure is highly relevant to these changes.

2. Fat Oxidation at the Air/Water Interface

The surface balance provides a possible means of monitoring the packing arrangement of the hydrocarbon chains of triglyceride monolayers. The spreading of triglyceride molecules occurs in a similar way on hydrophilic solids, so results from monolayers on water are therefore relevant to oxidation of fats on the surface of milk powder or the fat film covering the pore surface in cakes and cookies.

A study of soybean oil monolayers showed that oxidation gives significant effects on the Π-A isotherm when the triglyceride molecules, after spreading on the water surface, are left for more than half an hour exposed to air before compression (Moberger and Larsson, 1985). The two types of end product of the oxidation can even be identified from the appearance of the compression isotherms. Low molecular weight fission products (such as alcohols, aldehydes and ketones) show an increased pressure at high molecular area. When the liquid state is reached, they are squeezed out from the monolayer. The other type of oxidation product is cross-linked polymeric lipids. They give rise to a monolayer behaviour in which the pressure is not constant after the collapse point, but increases steadily. The reason is the continuous conformational changes of these amphiphilic polymers.

The monolayer compressibility can be used to follow the kinetics of oxidation. A logarithmic plot of the compressibility versus time is linear, and this is consistent with a first order reaction (Moberger and Larsson, 1985). When the compressibility is examined in the monolayer gas phase, the oxidation seems to be complete after about 2.5 hours. This work demonstrates how oxidation is accelerated when the monolayer expands from a condensed phase into a gas phase with exposed hydrocarbon chains. The effect of antioxidants in the subphase could also be followed by this technique.

3. *Synergism of Antioxidants in Microemulsions*

Antioxidants constitute an important group of food additives, not only for sensory improvements of foods, but also for nutritional reasons, as will be discussed in the next section. General aspects of food antioxidants have been discussed elsewhere (*cf.* Coppen, 1990).

The function of an antioxidant in a food product is related to the physical location of the reacting molecules. A model system which clearly demonstrates the relationship between antioxidant effect and structure was obtained by forming an L2-phase (or microemulsion) with tocopherol in the oil and ascorbic acid in the water regions (Moberger *et al.*, 1987). The system was designed to simulate the coupled antioxidant effects of tocopherol and ascorbic acid in the cell membrane, as described in the introduction to this section.

In order to obtain a microemulsion consisting of edible components, the ternary system triglyceride oil/water/monoglycerides described in Chapter 3 was applied. The oxidation of soybean oil stored in open glass vessels at 40°C was followed. The oil itself was polymerized into a semi-solid mass after 100 days, whereas the corresponding microemulsion with 200 ppm α-tocopherol and 500 ppm ascorbic acid showed no detectable oxidation during the same time. However, a complication in the technical application of these results is the need for a high proportion of monoglycerides in order to form the microemulsion. It seems likely that other polar lipids may be found that are more effective in forming L2-phases at low concentrations. Then the synergism of C- and E-vitamins as antioxidants opens many technical possibilities to protect fats and oils in food.

One application of this mechanism, with the intention of protection of fish oil, has been examined (Jakobsson and Sivik, 1994). α-Tocopherol was added to the oil at a concentration of 600 ppm, and glycerol was used as solvent for ascorbic acid (5% (w/w)). The microemulsion was formed from 77% (w/w) fish oil, 20% monoglyceride and 3% glycerol solution. The oxidation (measured as anisidine and peroxide values) at 40°C was negligible even after one week, when the oil was organized in the microemulsion state. The corresponding microemulsion with water instead of glycerol showed a much faster rate of oxidation.

Phospholipids exhibit an antioxidant effect at elevated temperatures, that is due to the development of browning products which are antioxidants (Husain *et al.*, 1986). The surface activity of phospholipids indicates that this effect is especially important at oil/water interfaces.

G. NUTRITIONAL ASPECTS

If we consider the different food components in human nutrition, lipids have been the subject of a high proportion of the research. This is not surprising. Lipid metabolism plays a major role in arterial heart diseases.

During recent years, it has become clear that lipids are also involved in the development of many forms of cancer, *e.g.* that of the colon. Lipid nutrition is outside the scope of this book, but there are certain aspects related to molecular organization and physical properties that are relevant.

1. Fats and Oils

The mechanism behind low-caloric fats, like "caprinin" discussed above, is the reduced absorption through the intestinal wall as a consequence of the phase behaviour of the lipolysis products. Fat products with a high melting point, such as the fully-hardened fats used in continuous bread-making processes, can probably not be digested and absorbed in the intestines. This field offers new possibilities for low-caloric products based on fats and oils.

2. Food Emulsifiers

The possibilities for physical damage by strongly polar lipids in the gastrointestinal system were mentioned in Chapter 12. As cell membranes should be expected to be most sensitive, a simple model has been used to examine food emulsifiers in this context (Larsson and Johansson, 1978). The haemolytic effect when erythrocytes were exposed to various emulsifiers was measured. When compared with lysophosphatidylcholine, which is considered to be able to damage the gastric wall in concentrations above 100 μg/ml, sodium stearoyl-lactyl-lactate (NaSLL) is the only food emulsifier with a similar strong haemolytic effect. Emulsifiers like monoglycerides, PC, tetraglycerol esters and diacetyl tartaric acid esters (DATE) showed very weak effects. This study may be taken as a warning not to use NaSLL in food applications other than those involving starch, where the possibly harmful effect can be eliminated by the complexing reaction with amylose. In this connection, it should be mentioned that polysorbate 80 has been reported to cause inflammation and even ulcers in the stomach in mice (*cf.* Vulfson, 1993).

Enzymatic synthesis of food emulsifiers has been introduced recently, and a strong motivation is the mild reaction conditions with reduced risks for formation of undesirable side products (*cf.* Vulfson, 1993). Those prepared enzymatically may also be more effective. As an example, it can be mentioned that current work in our department has shown that 2-monoglycerides are more effective in starch complexing reactions than are the corresponding 1-isomers.

3. Oxidation of Lipids, Cholesterol and Cardiovascular Disease

Dietary interventions in order to inhibit atherosclerosis and reduce the risk of heart disease is a controversial subject. The recent findings that monounsaturated fatty acids are as effective as polyunsaturated ones in lowering LDL-cholesterol has changed dietary advice in favour of fats rich in oleic acid. Too high levels of linoleic acid suppress the immune system and increase the risk of gallstones (*cf.* Barnes, 1990). The ideal balance between dietary monounsaturated fatty acids and polyunsaturated ones of the *n*-3 and *n*-6 series is at present a field of intensive research. Another question, where recent knowledge has changed the basis for dietary advice, is the physiological effect of oxidized lipids. Kubow (1990) has reviewed the toxic effects of lipid peroxidation products. Special interest has recently been focused on cholesterol oxides and their role in cardiovascular disease. As this field is related directly to lipid oxidation and effects due to molecular organization, it will be considered here.

The protection of dietary fat against oxidation has now become a more important issue than the proportions of saturated fats and cholesterol in the food. The reason is that oxidative changes of LDL have been identified as the most significant factor in the formation of the atherosclerotic plaque (Ross, 1993). Oxidized LDL (oxLDL) is thus a key component in the development of lesions of atherosclerosis. Adherence of monocytes, induced by oxLDL at the endothelial wall, leads to migration of monocytes through the wall. The resulting macrophages become large "foam cells", due to lipid accumulation, and they form so-called fatty streaks and ultimately a fibrous plaque. It is important to understand the role of tocopherol, ascorbic acid, β-carotene, ubiquinone and glutathione/glutathione peroxidase in the protection against lipid oxidation *in vivo.*

OxLDL can probably be induced by oxidized lipids already present in the diet, and oxidation products of cholesterol might be the reason behind links found earlier between cholesterol and formation of atherosclerotic lesions. Oxidized lipids in LDL might also be formed *in vivo* by lipoxygenase activity, or by oxygen or hydroxyl radicals. The oxidation of cholesterol *in vivo* is now considered to be a significant risk factor in this connection (*cf.* Fitch, 1993). Such cholesterol oxidation can be expected to be initiated by free radicals, for example by hydrogen peroxide and superoxide radicals produced by leukocytes. This means that the location of cholesterol in relation to the oil/water interface is a relevant factor.

Oils of animal origin contain dissolved cholesterol. When such an oil is exposed to water, the cholesterol will form a monolayer at the interface. Any phospholipids present will also occur in the film, but for simplicity we will only discuss cholesterol here. The external water phase will also take up some cholesterol. The reason is that at equilibrium between a monolayer and water, the water phase is saturated with regard to cholesterol at the solubility of monomers, *i.e.* at about 10^{-6} M (Larsson, 1986). The presence of water therefore means that cholesterol dissolved in the oil will be accumulated at the interface, and it will be exposed to free radical reactions in the water phase.

Numerous oxidation products have been identified, and 25-

hydroxycholesterol and cholesterol epoxides are considered to be the most atherogenic oxides (Smith, 1987). According to the new knowledge in this field, it seems that protection against oxidation of foods containing cholesterol is more important than the reduction of dietary cholesterol.

H. FAT REPLACERS

Oils and fats have numerous functions in foods, and there is no magic formula by which they can be replaced without changing some of the properties at least. The consumer demand for low-fat and low-caloric foods has resulted in an increasing market for products described as fat substitutes or replacers. The present state of the art will be summarized below, and future trends in this field will be discussed.

A synthetic fat substitute consisting of sucrose polyesters with zero caloric value was the first concept in this field (introduced by Procter and Gamble). In spite of excellent scientific and technological work, it now seems fair to state that this idea has failed. The reason is their gastrointestinal side-effects and nutritional aspects (such as interference with the absorption of lipophilic vitamins).

Polar lipids or emulsifiers are used in low-fat foods in order to simulate the functional properties of the original product. Lipid-water phases can exhibit properties like those of fats and oils, for example forming dispersed particles which can dissolve lipophilic aromas. Emulsifiers can therefore replace fats and oils in foods merely by their bulk properties. Because of the content of "inner" water, such a phase will always have a lower caloric value per volume unit. The interfacial properties of the emulsifiers will usually contribute positively to the food texture also, *e.g.* by improved aeration or a more monodisperse emulsion structure.

Polysaccharides or hydrophobic protein aggregates can sometimes simulate the bulk flow properties of fats and oil dispersions. Well-known examples are "Simpless", prepared from whey, "N-oil" or "N-lite" prepared from starch, and microcrystalline cellulose "Avicel". These types of fat replacers have been classified as thickening agents, soluble bulking agents and microparticulates respectively (Glicksman, 1991).

Gelling agents are often used to stabilize emulsions, when the fat level is reduced. One example is w/o-emulsions (spreads) with much lower water content than margarine. Gelatine is frequently used today in such low-fat spread products.

Future developments in this field will certainly utilize new types of lipid-water phases. The L3-phase described in Chapter 3 demonstrates the possibility of supplying an oil which has about one tenth of the caloric value of a triglyceride oil. When the functional properties of plastic fats are needed, it will certainly be possible to reduce the solid fat content by producing fat crystals of optimal shape. The use of diglyceride crystals instead of triglycerides might be one possible route, as discussed in Chapter 2.

I. ENCAPSULATION OF LIPIDS

In order to retard oxidation and maintain flavours dissolved in lipids during storage, technology has been developed involving encapsulation of the lipids in a dried matrix of polysaccharides or of proteins (Matsuno and Adachi, 1993). Medium-chain fatty acid triglycerides (MCTs) are of special interest, because of their stability against oxidation. They are also absorbed differently compared to long-chain triglycerides, and this might offer physiological advantages in medical and nutritional applications.

Two methods have been developed. The one used most commonly is to apply spray-drying to an appropriate o/w-emulsion containing the polysaccharide or protein in the aqueous phase. The other method is based on the solubility of wheat storage proteins in ethanol. For example, linoleic acid can be solubilized with gliadin in ethanol, and on evaporation of the ethanol the fatty acid becomes encapsulated by a gliadin film.

CHAPTER 15

FUTURE ASPECTS OF FUNCTIONAL APPLICATIONS OF LIPIDS

Some general trends in the development of new technology and on future applications of lipids have been indicated in the various preceding chapters. Certain significant aspects related to the future of lipid functionality will be considered below.

A. New Oilseed Crops

Traditional plant breeding, based on manual selection, has given us the major oil crops we know today. A good example is rapeseed. During the 1970s, the content of erucic acid was successfully eliminated and replaced mainly by oleic acid for nutritional reasons. A disadvantage with this new oil turned out to be its functional character, however. When it was hardened for use in products like margarine, it gave β'-crystals with too short a lifetime (*cf.* Chapter 14). It seemed probable that the reason for the rapid $\beta' \rightarrow \beta$ transition was that the oil was too homogeneous in the distribution of chain-lengths of the fatty acids (>80% C_{18}). One way to overcome this was additional plant breeding toward a higher relative content of C_{16}-fatty acids. As expected, this oil turned out to be close to ideal.

Mutations can be achieved by chemicals or by radiation. Sometimes such modified seeds may still be able to grow, and the new plant can then be screened for a certain desired property (for example, a reduced linolenic acid content). A recent application of this technique was reported from CSIRO in Canberra, where about 50% of linolenic acid in linseed was replaced by linoleic (Green, 1988).

The revolutionary technique of gene transfer has not yet had much impact on the production of oils in plants. There are several reasons for this; the most important one is that the complete pathway for triglyceride biosynthesis is not known (*cf.* Taylor, 1991). Once this is fully understood it seems likely that we will see completely new oil crops in future agriculture. Of about 75,000 plant species in the world estimated to be edible, only a few are used for food production (about 20 species provide 90% of the world food supply). This is especially true when we consider oil crops.

Some interesting new crops for temperate climate zones have recently been characterized (Muuse *et al.*, 1991). Oil from *Limnanthes alba* contains fatty acids of long chain-length, and it was found to be very stable against oxidation. *Euphorbia lagacae* and *Dimorphoteca pluvialis* contain fatty acids with epoxy- and hydroxydiene groups. These reactive fatty acids are interesting oleochemical raw materials. Two alternative crops to rapeseed with high erucic acid content are also described.

The South American flowering plant family, *Cuphea*, has unique possibilities as an industrial crop because of its content of fatty acids of medium chain-length (Graham and Kleiman, 1992). Certain species produce equal proportions of caprylic, capric, lauric and myristic acids.

Microalgae are proposed to be a source for new special lipids over the next ten years (Kyle, 1992). Microalgae or phytoplankton are the major source of eicosapentaenoic and docosahexaenoic acids in the food chain. There are thousands of different species, and they can be cultivated in bioreactors or ponds. About 10g oil/m^2/day can be produced, and β-carotene is already produced commercially in open ponds. Isotopically labelled lipids can be produced by simply supplying an isotopic precursor in the culture medium. Deuterated lubricants and lipids for NMR medical imaging are examples of such products. Specially designed lipids, for example for infant formulas, will probably be based on microalgae production in the future.

B. OIL PROCESSING

New techniques of oil extraction involve the use of enzymes to open up the biological tissue and make the lipids available for extraction. The use of supercritical carbon dioxide was discussed in Chapter 2. Whether or not the use of this extraction medium will be economically feasible in the future is hard to predict at present. In any case, this technique will certainly be applied in fractionation.

The crude oil obtained after extraction contains phospholipids, which are removed by so-called degumming. Classical degumming is done by the addition of water (at 60-90°C) and is based on the phase equilibrium of the phospholipid/triglyceride/water system. Thus an aqueous phospholipid phase is formed at the bottom, and this can be removed easily by centrifugation. Crude soybean oil contains 2-3% phospholipids, which in this way can be reduced to less than 1%. New methods ("super-degumming") involve use of citric acid, and allow a reduction of the phospholipids down to about 30 ppm (Trommelen, 1993). The kinetics of the formation of the aqueous phases are related to the proportions of PC/PE/PI and especially the presence of calcium and magnesium ions.

The development of heat-resistant separation membranes has made possible the separation of phospholipids by ultrafiltration of the crude oil. When the oil is dissolved in hexane, these polar lipids will form reversed micelles, which are large enough to be separated by such membranes (Krishnamurthy, 1992).

The introduction of enzyme technology offers new possibilities. Phospholipase C will convert phospholipids into diglycerides and water-soluble phosphates (Graille, 1993). A good source for phospholipase C is *Bacillus cereus*, and future production can be expected for this application at a reasonable price. Neutralization and bleaching of oils based on such enzyme technology are even being considered now.

The problem of *trans* fatty acid formation during the hydrogenation of oil is under critical evaluation at present (*cf.* Berger, 1993). Thus, there is a desire to develop new processes which can reduce the amount of *trans*-fatty acids in hardened fats. By interesterification of fats containing saturated fatty acids with oils rich in unsaturated acids, such fats can be obtained. In order to modify fat crystallization as desired, recent work has shown that the combination of full hydrogenation and enzymatic interesterification is a powerful way to tailor a fat, free from *trans* fatty acids, according to functional needs (Mohamed and Larsson, 1994). A study of a solid fat for margarine, prepared by interesterification of cottonseed oil and fully hydrogenated soybean oil by lipases of different specificity, showed that only at a certain degree of positional specificity of the lipase, could an acceptable crystallization behaviour be achieved.

C. ENZYME TECHNOLOGY

The use of lipases and phospholipases for lipid synthesis and modification is a highly active field of research at present. Technological applications have been made possible through the cheap production of enzymes from microorganisms.

Lipases can have a very broad substrate specificity, and they are extremely stable when exposed to different chemical environments. The first crystal structures of the lipase from *Rhizomucor miehei* (Brady, 1990) and from human pancreas (Winkler *et al.*, 1990) revealed how the active site is covered by an α-helix surface loop. This loop is folded back, exposing the active site, when the enzyme is oriented at an oil/water interface, or when it is kept in a non-polar solvent. The fact that lipases can maintain their activity in organic solvents is the basis for ester synthesis. As the reaction produces water, removal of water will drive the reaction equilibrium towards ester synthesis. Different lipases are now available with varying selectivities in relation to lipid class, fatty acid and position (*cf.* Bloomer, 1992). Reactions such as transesterification of oil mixtures are easy to carry out from a technological point of view; the enzyme adsorbed on mineral particles is added directly to the oil mixture.

Phospholipids can also be modified by lipases (Svensson *et al.*, 1993). In order to utilize an interface for this reaction, glyceride synthesis has been performed in a foam (Oh *et al.*, 1992).

There is clearly an enormous potential in enzymatic synthesis and modification of lipids, and this development has not even started yet. Tailoring of fats and polar lipids according to functional requirements will be possible, and a knowledge of relationships between chemical structure, molecular organization and physical properties is a pre-requisite for such new technology. With increasing knowledge of the nutritional benefits of specific fatty acids in the 2-position of triglycerides, there will be possibilities to produce the desired positional isomers. Polar lipids with new functions can also be expected; the L3-phase formed by $C_6C_{16}PC$ (Chapter 3) is an illustrative example (Svensson *et al.*, 1993).

D. BIOSURFACTANTS

Surfactants can be regarded as synthetic analogues to polar lipids that are capable of forming micellar (L1) solutions. At present, the principal raw material for their synthesis is petroleum. Certain microorganisms can produce such polar lipids, biosurfactants, which can probably compete with petrochemical surfactants in the future. The main advantage of biosurfactants is their biodegradable character.

One group of biosurfactants is glucolipids (*cf.* Fiechter, 1992). Rhamnolipids from *Pseudomonas aeruginosa* and *Pseudomonas fluorescens* are examples of such glycolipids. Their physiological role is to solubilize the oil and make it available as a substrate for growth of the organism.

A powerful biosurfactant called surfactin consists of a cyclic amino acid structure (Glu-Leu-Leu-Val-Asp-Leu-Leu) linked to 3-hydroxy-13-methyl-tetradecanoic acid (*cf.* Fiechter, 1992).

These lipids are also of interest because of their antimicrobial properties. Synthetic analogues are simple amino acid esters, for example of lysine or glycine, with long-chain alcohols (Lindstedt *et al.*, 1993). This antimicrobial surfactant is hydrolysed slowly in water solution, and it might therefore be used as an antimicrobial agent with a temporary effect.

There are also polymeric surface-active molecules produced by microorganisms that are defined as biosurfactants. The best known is "emulsan", the only biosurfactant on the market at present.

If phospholipids are hydrolysed into lyso-phospholipids, the resultant polar lipids have surfactant properties and should be classified as biosurfactants. Enzymatic synthesis based on cheap vegetable phospholipids should be competitive in comparison with other biosurfactants in the future. The possibility of genetic engineering towards molecules with optimal performance suggests that in the future biosurfactants will successively take over many applications where petrochemical surfactants are used today (*cf.* Kosaric *et al.*, 1993).

Oligomerization of fatty acids has recently been developed into an industrial process based on radical reactions of acetic acid (Fristad, 1992). Large amphiphilic molecules with physical properties suitable for lubrication and emulsion stabilization can be produced in this way.

The oleosins are proteins with special stabilizing properties for the oil/water interface, and they have therefore been proposed for use as future emulsifiers. According to their general surface-active properties, they seem to resemble β-caseins (see Figure 8.3) (Murphy, 1993).

E. TECHNOLOGY BASED ON HIGH PRESSURE AND SUPERCRITICAL CARBON DIOXIDE

The thermodynamic basis for pressure-induced effects in lipid systems is Le Chatelier's principle: "It is a feature of stable equilibrium that spontaneous processes induced by deviation from the initial state proceed in such a direction as to restore the system to the initial state". This means that an increase in pressure applied to a liquid triglyceride will move the equilibrium between the solid and liquid state towards the solid, as the solid has a smaller volume (crystallization of a closed system will result in reduced pressure).

Furthermore, the most stable polymorphic form of a fat has the highest density, and applied pressure will therefore drive the equilibrium towards the most stable crystal form. The application potential of pressure in the processing of fats was discussed in Chapter 2. There are additional possibilities to influence lipid phase transitions in aqueous systems, however. If we first consider the simple case of an o/w-emulsion, the application of high pressure means that the water to ice transition is shifted to a lower temperature (as ice has a lower density than water). Pressure-induced crystallization or phase transition of the fat can therefore be performed at temperatures below 0°C without destroying the emulsion. The situation becomes more complex in aqueous phases of polar lipids, as the water molecules adopt structures different from those in bulk water. Basic studies are needed in order to be able to predict pressure-effects on phase transitions in lipid-water systems. It seems probable that there will be new possibilities for applications based on such knowledge. A hypothetical example might be encapsulation of enzymes into the cubic lipid-water phase for stabilization and protection of the protein conformation. The release of the enzymes in a bioreactor can then be achieved instantaneously by a pressure-induced phase transition from cubic to L_α

When supercritical carbon dioxide (s-CO_2) is used for separation or as a medium for enzymatic synthesis of lipids, it is important to understand the effect of pressure on lipid phase properties. Pressure-induced crystallization can be achieved in the s-CO_2 medium, and this might result in separation of particular lipid species. In addition, this mechanism could probably be utilized to drive enzymatic reactions, for example interesterification by lipases, if the desired triglyceride separates from the reaction sites.

Physiological effects of s-CO_2 through interactions with lipids have been neglected. In connection with work on s-CO_2 at our department, it struck me that carbon dioxide will become supercritical at ocean depths below about 750 m in animals with a body temperature above 31°C. This might be utilized by whales as a mechanism to provide a temporary sink for carbon dioxide produced metabolically.

The toothed whales are amazing divers; the master is the sperm whale (*Physeter catodon*), which is able to dive to a depth of more than one km and stay submerged for an hour. All toothed whales have two fat depots in the forehead, one is the "melon" (or "junk" in the sperm whale) and the other is the mandibular fat body (Blomberg, 1978). The lipids in these head-oils differ between species of toothed whales. Wax esters and triglycerides of low molecular weight are most abundant, and this results in an oil with low viscosity (Litchfield and Greenberg, 1974). The functional role of the head fat depots remains unknown. It has been proposed that the "melon" is involved in sound processing, and functions as an acoustic lens (Blomberg, 1978). Pressure-induced crystallization to regulate the density has also been proposed (see Chapter 2). The sperm whale, the best diver, has a fat depot of an enormous size.

At the body temperature of the whale and at a water depth below 740 m, the carbon dioxide becomes supercritical. Its partition between aqueous and oil media, such as blood and the fat depot in the forehead, is then shifted drastically. This means that the carbon dioxide produced by the muscular glucolysis can accumulate in the head-oil phase during the stay at a high depth. When the carbon dioxide is released during the rapid swim to the surface, there is only a short time during which it can be eliminated via the lungs.

This wild speculation seem a natural end to this book, which started with exact results from single-crystal analysis and successively moved into more and more complex functions of lipids in biology and technology, where some speculations are often necessary in discussions of functionality.

REFERENCES

Abrahamsson, J., Abrahamsson, S., Hellqvist, B., Larsson, K., Pascher, I. & Sundell, S. *Chem. Phys. Lipids* **19**, 213 (1977).

Abrahamsson, S., Dahlen, B. & Pascher, I. *Acta Cryst.* **B33**, 2008 (1977).

Abrahamsson, S., Pascher, I., Larsson, K. & Karlsson, K.-A. *Chem. Phys. Lipids* **8**, 152 (1972).

Abrahamsson, S. & Ryderstedt-Nahringbauer, I. *Acta Cryst.* **15**, 1261 (1962).

Al'Rajab, A., Ahrén, B., Rozga, J. & Bengmark, S. *J. Surg. Research* **50**, 212 (1991).

Albrecht, O., Gruler, H. & Sackmann, E. *J. Colloid Interface Sci.* **79**, 319 (1981).

Alving, C.R. & Richards, R.L. *Immunol. Lett.* **40**, 359 (1986).

Andersson, D., Wennerström, H. & Olsson, U.J. *J. Phys. Chem.* **93**, 4243 (1989).

Andersson, J.W. & Budges, S. *Am. J. Clin. Nutr.* **35**, 840 (1982).

Andersson, S., Blum, Z., Hyde, S., Landt, T., Larsson, S., Lidin, S. & Ninham, B. *Language of Shape* (Elsevier, Amsterdam, 1994 (in the press)).

Andersson, S., Hyde, S.T., Larsson, K. & Lidin, S. *Chem. Rev.* **88**, 221 (1988).

Annesini, M.C., Finazzi-Agrà, A. & Mossa, G. *Chimica Oggi* **10**, 11 (1992).

Arishima, T., Sago, N., Mori, H. & Sato, K. *J. Am. Oil. Chem. Soc.* **68**, 710 (1991).

Aronshime, J., Sarig, S. & Garti, N. *Food Structure* **9**, 337 (1990).

Arvidsson, G., Brentel, I., Kahn, A., Lindblom, G. & Fontell, K. *Eur. J. Biochem.* **152**, 753 (1985).

Bailey, A.E. *Melting and Solidification of Fats* (Wiley Interscience, New York, 1950).

Barnes, P. *Lipid Technol.* **2**, 3 (1990).

Barker, N. & Hadgraft, J. *Int. J. Pharmac.* **8**, 202 (1981).

Barla, P., Larsson, K., Ljusberg-Wahren, H., Norin, T. & Roberts, K. *J. Sci. Food Agric.* **30**, 864 (1979).

Behan-Martin, M.K., Jones, G.R., Bowler, K. & Cossins, A.R. *Biochim. Biophys. Acta* **1151**, 216 (1993).

Beissinger, R.L., Farmer, M.C. & Gossage, J.L. *Am. Soc. Artif. Intern. Organs* **32**, 58 (1986).

Berger, K. in *Food Emulsions* (eds. Larsson, K. & Friberg, S.) p. 141 (Marcel Dekker, New York, 1989).

Berger, K. *Lipid Technol.* **5**, 37 (1993).

Bhaskar, K.R., Gavik, P., Turner, B.S., Bradley, J.D., Bausil, R., Stanley, H.E. & LaMout, J.E. *Nature* **360**, 458 (1992).

Biliaderis, C.G. & Galloway, G. *Carbohydrate Res.* **189**, 31 (1989).

Birker, P.J.M.W.L., Jong, S.D., Roijers, S., Roigers, E.C. & Soest, T.C.v. *J. Am. Oil Chem. Soc* **68**, 895 (1991).

Blodgett, K.B. *J. Am. Chem. Soc.* **57**, 1007 (1935).

Blomberg, J. *Prog. Chem. Fats other Lipids* **16**, 257 (1978).

Bloomer, S. *Lipase-catalyzed Lipid Modification in Non-aqueous Media* (Thesis, University of Lund, Sweden, 1992).

Blute, I., Holmberg, K., Jansson, M., Oh, S.G. & Shah, D.O. *J. Am. Oil Chem. Soc.*, in the press (1994).

Bogren, H. & Larsson, K. *Biochim. Biophys. Acta* **75**, 65 (1963).

Borman, S. *Chem. & Eng. News* **70**, 57 (1992).

Bowstra, J.A., de Vries, M.A., Gooris, G.S., Bras, W., Brusse, J. & Ponec, M. *J. Controlled Release* **15**, 209 (1991).

Brady, L., Brzozowski,A.M., Derewenda, Z.S., Dodson, E., Dodson, G., Tolley, S.,

Turkenburg, J.P., Christiansen, L., Huge-Jensen, B., Norskov, L., Thim. L. & Menge, U. *Nature* **343**, 767 (1990).
Brocklehurst, T., Parker, M., Gunning, P. & Robins, M. *Lipid Technol.* **4**, 83 (1993).
Buchheim, W. & El-Nour, A.M.A. *Fat Sci. Technol.* **10**, 369 (1992).
Buchheim, W. & Larsson, K. *J. Colloid Interface Sci.* **117**, 582 (1987).
Buettner, G.R. *Arch. Biochem. Biophys.* **300**, 535 (1993).
Bundgaard, M.J. *Ultrastruct. Res.* 88, 1 (1984).
Bundgaard, M.J., FrØkjaer-Jensen, J. & Crone, C. *Proc. Natl. Acad. Sci. USA* **76**, 6439 (1979).
Bungenberg de Jong, H.G. in *Colloid Interface Science* (ed. Kruyt, H.R.) p. 232 (Elsevier, New York, 1949).
Burg, B. & Jeschke, R. *Fat Sci. Technol.* **94**, 249 (1992).
Bursh, T., Larsson, K. & Lundquist, M. *Chem. Phys. Lipids* **2**, 102 (1968).
Busch, G. & Neuwald, F. in *Proc. VII Congress IFSCC* (ed. Proppe, A.) p. 171 (Gesellschaft Deutsches Kosmetik-Chemiker, Hamburg, 1973).
Butler, B.D., Lichtenberg, L.M. & Hills, B.A. *Am. J. Physiol.* **244**, 645 (1983).
Caffrey, M. *Trends Anal. Chem.* **10**, 156 (1991).
Carlsson, A., Karlström, G. & Lindman, B. *Colloids and Surfaces* **47**, 147 (1990).
Carlsson, T., Larsson, K. & Miezis, Y. *Cereal Chem.* **55**, 168 (1978).
Carlsson, T.L.-G., Larsson, K., Dinh-Nguyen, N. & Krog, N. *Starch* **31**, 222 (1979).
Carrigan, P. & Bates, T. *J. Pharm. Sci.* **62**, 1476 (1973).
Cebula, D.J., Clemments, D.J.M., Povey, M.J.W. & Smith, P.R. *J. Am. Oil Chem. Soc.* **69**, 130 (1992).
Cevc, G. *Chem. Phys. Lipids* **57**, 293 (1991)
Chaganaud, P., Jenkinson, H.F. & Tannock, G.W. *Microb. Ecology Health Disease* **5**, 121 (1992).
Chang, F., Schwarzendruber, D.C., Wertz, P.W. & Squier, S. *Biochim. Biophys. Acta* **1150**, 98 (1993).
Chapman, D. *Chem. Rev.* **62**, 433 (1962).
Chapman, D. & Charles, S.A. *Chem. Britain* **28**, 253 (1992).
Chawla, P., Man, J.M.D. & Smith, A.K. *Food Microstructure* **9**, 329 (1990).
Chen, Q., Sternby, B. & Nilsson, Å. *Biochim. Biophys. Acta* **1004**, 37288(1989).
Christensen, O.B., Holst, R., Hradil, E. & Brolund, L. *Acta Derm. Venereol.* **146**, 96 (1989).
Christie, W.W. *High-Performance Liquid Chromatography and Lipids* (Pergamon Books, Oxford, 1987).
Clarkson, C.E.A.M. *J. Chem. Soc.* 666 (1934).
Cooper, E.R. *J. Pharm. Sci.* **73**, 1153 (1984).
Copper, P. *Lipid Technol.* **2**, 95 (1990).
Craig, S.J., Ninham, B.W. & Pashley, R.M. To be published (1994).
Craven, B.M. *Nature* **260**, 727 (1976).
Craven, B.M. & Guerina, G. *Chem. Phys. Lipids* **24**, 157 (1979).
Craven, B.M. & Tita, G.T.D. *J. Chem. Soc. Perkin II* 814 (1976).
Curatolo, W., Small, D.M. & Shipley, G.G. *Biochim. Biophys. Acta* **168**, 11 (1977).
Danielsson, I. & Lindman, B. *Colloids and Surfaces* **3**, 391 (1981).
Dardel, O.v., Mebius, C. & Mossberg, T. *Opusc. Med.* **20**, 255 (1973).
Davenport, H.W. *Gastroenterology* **59**, 505 (1970).
Davier, S.S., Hadgraft, J. & Palin, K.Y. in *Encyclopedia of Emulsion Technology* (ed. Becher, P.) p. 159 (Marcel Dekker, New York, 1985).
Deckelbaum, R.J., Hamilton, J.A., Moser, A., Bengtsson-Olivecrona, G., Buthul, E., Carpentier, Y.C., Gutman, A. and Olivecrona, T. *Biochemistry* **29**, 1136 (1990).
Dennis, E.A. *Drug Development Res.* **10**, 205 (1987).
Dervichian, D.G. *J. Chem. Phys.* **7**, 931 (1939).
Dickenson, E.I. in *Emulsions - A Fundamental and Practical Approach* (ed. Sjöblom, J.) p. 23 (Kluwer Academic Publ., London, 1992).
Doyne, T.H. & Gordon, J.T. *J. Am. Oil Chem. Soc.* **45**, 333 (1968).
Dunjic, B.S., Axelsson, J., Ar'Rajab, A., K. Larsson, K. & Bengmark, S. *Scand. J. Gastroenterology* **28**, 89 (1993).
Elder, M., Hitchcock, P., Mason, R. & Shipley, G.G. *Proc. Roy. Soc.* **A354**, 157 (1977).
Elias, P.M. *J. Invest. Dermatol.* **80**, 40 (1983).
Elias, P.M. *J. Controlled Release* **15**, 199 (1991).

Elias, P.M., Gophinatan, M.D. & Menon, K. *J. Invest. Dermatol.* **91**, 3 (1988).
Eliasson, A.-C. & Larsson, K. *Cereals in Breadmaking. A Molecular/Collodial Approach* (Marcel Dekker, New York, 1993).
Eliasson, A.-C., Larsson, K., Andersson, S., Hyde, S.T., Nesper, R. & Von Schnering, H.-G. *Starch* **39**, 147 (1987).
Elson, L.A., Mitchley, B.C.V., Collings, A.J. & Schneider, R. *Eur. J. Clin. Biochem. Res.* **15**, 87 (1970).
Engblom, J. & Engström, S. *Int. J. Pharmac.*, in the press (1994).
Engel, R.H. & Fahrenbach, M.J. *Proc. Soc. Exp. Med.* **129**, 772 (1968).
Engström, S. *Fat Sci. Technol.* **94**, 173 (1992).
Engström, S. To be published (1994).
Engström, S., Larsson, K. & Lindman, B. *English Patent* 314 689, (1987).
Engström, S., Larsson, K. & Lindman, B. *U.S. Patent* 5.151272, (1992).
Ericsson, B. & Hegg, P.O. *Prog. Colloid Polymer Sci.* **70**, 92 (1985).
Ericsson, B., Larsson, K. & Fontell, K. *Biochim. Biophys. Acta* **729**, 23 (1983).
Ernst, J., Sheldrik, W.S. & Fuhrhop, J.H. *Z. Naturforsch.* **34B**, 706 (1979).
Fahey, D.A. & Small, D.M. *Biochemistry* **25**, 4468 (1986).
Fahey, D.A. & Small, D.M. *Langmuir* **4**, 589 (1988).
Faiba, R., Ar'Rajab, A., Willén, R., Andersson, R., Ahrén, B., Larsson, K. & Bengmark, S. *Digestion* **53**, 35 (1992).
Farren, S.B., Hope, M.J. & Cullis, P.R. *Biochem. Biophys. Res. Commun.* **111**, 675 (1983).
Fiechter, A. *Lipid Technol.* **3**, 286 (1993).
Fiorentini, D., Landi, L., Barzanti, V. & Cabrini, L. *Free Rad. Res. Commun.* **6**, 243 (1989).
Fitch, B. *INFORM* **4**, 800 (1993).
Forslind, B. *Acta Derm. Scand.* in the press (1994).
Friberg, S.E., Mandell, L. & Larsson, M. *J. Colloid Interface Sci.* **29**, 155 1969).
Friend, D.R. & Heller, J. in *Pharmaceutical Skin Penetration Enhancement* (eds. Walters, K.A. & Hadgraft, J.) p. 31 (Marcel Dekker, New York, 1993).
Fristad, W. *Henkel-Referate* **28**, 39 (1992).
Fuller, R. *Gut* **32**, 439 (1991).
Gargouri, Y., Morean, H. & Verger, R. *Biochim. Biophys. Acta* **1006**, 255 (1989).
Garside, J. in *Food Structure and Behaviour* (ed. Garside, J.) pp. 1-35 (Academic Press, London, 1987).
Garti, N. & Sato, K. *Crystallization and Polymorphism of Fats and Fatty Acids* (Marcel Dekker, New York, 1988).
Glantz, P. & Larsson, K. *J. Dispersion Sci. Technol.* **3**, 573 (1982).
Glantz, P., Larsson, K. & Nyquist, G. *Odont. Rev.* **27**, 265 (1976).
Glicksman, M. *Food Technol.* **45**, 94 (1991).
Goggin, P.M., Northfield, T.C. & Spyckel, R.T. *Scand. J. Gastroenterol.* **26**, 65 (1991).
Goldstein, J.L. & Brown, M.S. *Annu. Rev. Biochem.* **46**, 897 (1977).
Goto, M. & Asada, E. *Bull. Chem. Soc. Jpn.* **51**, 70 (1978).
Goto, M., Kodali, D.R., Small, D.M., Honda, K., Kozawa, K. & Uchida, T. *Proc. Natl. Acad. Sci. USA* **89**, 8083 (1992).
Graham, S.A. & Kleiman, R. *Ind. Crops and Products* **1**, 31 (1992).
Grahame, G.R., Torchia, M.G., Dankevich, K.A. & Ferguson, F.A. *Bull. Perit. Dial.* **5**, 109 (1985).
Graille, J. *Lipid Technol.* **5**, 11 (1993).
Green, A.G. in *Proceedings of the World Congress on Biochemistry for the Fats and Oil Industry* (ed. Applewhite, T.H.) p. 55 (AOCS, Champaign, Ill., 1988).
Greene, R.S., Downing, D.T., Pochi, P.E. & Strauss, J.S. *J. Invest. Dermatol.* **54**, 204 (1970).
Gregoriadis, G. *Immunology Today* **11**, 89 (1990).
Grool, F.G.v.D., Gonzales-Manas, J.M., Lakey, J.H. & Pattus, F. *Nature* **354**, 408 (1991).
Gudmundsson, M. *Cereal Starches* (Thesis, University of Lund, Sweden, 1992).
Gudmundsson, M., Svensson, E. & Eliasson, A.-C. To be published (1994).
Gulik-Krzywicki, T., Aggerbeck, L.P. & Larsson, K. in *Surfactants in Solutions* (eds. Mittel, L. & Lindman, B.) p. 237 (Plenum Press, New York, 1984).
Gulik-Krzywicki, T. & Larsson, K. *Chem. Phys. Lipids* **35**, 127 (1984).
Gulik-Krzywicki, T., Shechter, E., Luzzati, V. & Faure, M. *Nature* **223**, 1116 (1969).
Gunstone, F.D., Harwood, J.L. & Padley, F.B. (eds) *The Lipid Handbook* (Chapman and Hall, London, 1986).

Hadgraft, J., Williams, D.G. & Allan, G. in *Pharmaceutical Skin Penetration Enhancement* (eds. Walters, K.A. & Hadgraft, J.) p. 175 (Marcel Dekker, New York, 1993).
Hallman, M., Pohjavnori, M., Bry, K. & Meritt, T.A. *Int. J. Technol. Assessment in Health Care* **7**, 21 (1991).
Hamilton, J.A. *Biochemistry* **28**, 2514 (1989).
Hamilton, J.A. & Small, D.M. *Proc. Natl. Acad. Sci. USA* **78**, 6878 (1981).
Hamilton, J.A., Fujito, D.T. & Hammer, C.F. *Biochemistry* **30**, 2894 (1991).
Hamilton, J.A. *NIPS* 7, December issue (1992).
Harlos, K. & Eibl, H. *Biochemistry* **19**, 895 (1980).
Harlos, K., Eibl, H., Pascher, I. & Sundell, S. *Chem Phys. Lipids* **34**, 115 (1984).
Harwood, J.L., Cryer, A. & Gurr, M.I. in *The Lipid Handbook* (eds. Gunstone, F.D., Harwood, J.L. & Padley, F.B.) p. 527 (Chapman And Hall, London, 1986).
Hashida, M., Takahashi, Y., Muranishi, S. & Sezaki, H. *J. Pharmacokin. Biopharm.* **5**, 225 (1977).
Hauser, H., Pascher, I., Pearson, R.H. & Sundell, S. *Biochim. Biophys. Acta* **650**, 21 (1981).
Hayward, J.A. & Chapman, D. *Biomaterials* **5**, 135 (1984).
Hein, M., Madefessel, C., Haag, B., Teichmann, K., Post, A. & Galla, H.-J. *Chem. Phys. Lipids* **63**, 223 (1992).
Helfrich, W.Z. *Naturforsch.* **A28**, 693 (1973).
Hernqvist, L., Herslöf, B., Larsson, K. & Podlaha, O. *J. Sci. Food Agric.* **32**, 1197 (1981).
Hernqvist, L. & Larsson, K. *Fette Seifen Anstrichm.* **84**, 349 (1982).
Herrington, T.M., Midmore, B.R. & Sahi, S.S. *ACS Symposium Series* **448**, 82 (1991).
Hills, B.A. *The Biology of Surfactant* (Cambridge University Press, Cambridge, 1988).
Hills, B.A. *J. Proc. Royal Soc. New South Wales* **122**, 19 (1989a).
Hills, B.A. *J. Rheumatol.* **16**, 82 (1989b).
Hills, B.A. *Med. J. Australia* **153**, 76 (1990).
Hirose, M. *Trends Food Sci. Technol.* **4**, 48 (1993).
Hirsch, M., Gache, D. & Noske, W. *Cell Tissue Res.* **252**, 165 (1988).
Hope, M.J., Bally, M.B., Mayer, L.D., Janoff, A.S. & Cullis, P.R. *Chem. Phys. Lipids* **40**, 89 (1986).
Hope, M.J. & Cullis, P.R. *Biochem. Biophys. Res. Commun.* **21**, 846 (1980).
Hoppe, U. & Larsson, K. *J. Dispersion Sci. Technol.* **2**, 433 (1981).
Hoverstad, T. *Scand. J. Gastroenterol.* **21**, 257 (1986).
Husain, S.R., Terao, J. & Matsushita, S. *J. Am. Oil. Chem. Soc.* **63**, 1457 (1986).
Hybl, A. & Dorset, D. *Acta Cryst.* **B27**, 977 (1971).
Hyde, S.T., Ericsson, B., Andersson, S. & Larsson, K. *Z. Kristallogr.* **168**, 213 (1984).
Hyde, S.T., Andersson, S. & Larsson, K. *Z. Kristallogr.* **174**, 237 (1986).
Ishikara, K., Aragaki, R., Ueda, T., Watanabe, A. & Nakabayashi, N. *J. Biomed. Mat. Res.* **24**, 1069 (1990).
Israelachvili, J.N. *Intermolecular and Surface Forces* pp. 1-232 (Academic Press, London, 1985).
Israelachvili, J.N., Mitchell, D.J. & Ninham, B.W. *Biochim. Biophys. Acta* **470**, 185 (1977).
Israelachvili, J.N. & Pashley, R.M. *Nature* **300**, 341 (1982).
Israelachvili, J.N. & Wennerström, H. *Langmuir* **6**, 873 (1990).
Itsu, L.-Y. & Nordman, C.E. *Science* **220**, 604 (1983).
Jakobsson, M. & Sivik, B. *J. Dispersion Sci. Technol.* in the press (1994).
Janiak, M.J., Small, D.M. & Shipley, G.G. *Biochemistry* **15**, 4575 (1976).
Jensen, L.H. & Mabis, A.J. *Nature* **197**, 681 (1963).
Jewel, G.G. & Meara, M.L. *J. Am. Oil Chem. Soc.* **47**, 535 (1970).
Johansson, B.B. *Adv. Exp. Med. Biol.* **274**, 25 (1990).
Johansson, M.L., Molin, G., Jeppsson, B., Nobaek, S., Ahrné, S. & Bengmark, S. *Appl. Environm. Microbiol.* **59**, 15 (1993).
Johnson, A.G. & McDermott, S.J. *Gut* **15**, 710 (1974).
Johnson, K.A., Westermann-Clark, G.B. & Shah, D.O. *Pharmaceut. Res.* **6**, 239 (1989).
Jong, S.D. *Triacylglycerol Crystal Structure and Fatty Acid Conformation* (Thesis, University of Utrecht, Netherlands, 1980).
Juliano, R.L. & Stamp, D. *Biochem. Pharmacol.* **27**, 21 (1978).
Kamp, F. & Hamilton, J.A. *Proc. Natl. Acad. Sci. USA* **89**, 11367 (1992).
Kao, Y.-C. & Lichtenberger, L.M. *Gastroenterology* 92, 1459 (1987).
Katz, M. in *Pharmaceutics in Drug Design* (ed. Arien, E.J.) p. 97 (Academic Press, London,

1973).
Kawahara, Y. *INFORM* **4**, 663 (1993).
Kayali, I., Suhery, T., Friberg, S., Simion, F.A. & Rhein, L.D. *J. Pharmaceut. Sci.* **80**, 428 (1991).
Kekicheff, P. & Cabane, B. *Acta Cryst.* **B44**, 395 (1988).
Kellens, M., Meussen, W. & Reynaers, H. *Chem. Phys. Lipids* **55**, 163 (1990).
Kellens, M. & Reynaers, H. *Fat Sci. Technol.* **94**, 286 (1992).
Kernon, J. *Lipid Technology* **4**, 113 (1992).
Kikuchi, H., Yamanchi, H. & Hirota, S. *Chem. Pharm. Bull.* **39**, 1552 (1991).
Kim, H.R., Eliasson, A.-C. & Larsson, K. *Carbohydrate Polymers* **19**, 211 (1992).
Kirk, G.L., Gruner, S.M. & Stein, D.L. *Biochemistry* **23**, 1093 (1984).
Klascha, F. *Årtzliche Kosmetologie* **12**, 88 (1982).
Knoll, W., Schmidt, G., Rötzer, H., Henkel, T., Pfeiffer, W., Sackmann, E., Mittler-Neher, S. and Spinke, J.. *Chem. Phys. Lipids* **57**, 363 (1991).
Kodali, D.R., Atkinson, D. & Small, D. *J. Phys. Chem.* **93**, 4683 (1989).
Kosaric, N., Cairns, W.L. & Gray, N.C.C. *Biosurfactants and Biotechnology* (Marcel Dekker, New York, 1987).
Krishnamurthy, R.G. *INFORM* **3**, 1130 (1992).
Krog, N. & Larsson, K. *Chem. Phys. Lipids* **2**, 129 (1968).
Krog, N. & Larsson, K. *Fat Sci. Technol.* 94, 55 (1992).
Krog, N. & Lauridsen, *J. in Food Emulsions* (ed. Friberg, S.) p. 67 (Marcel Dekker, New York, 1976).
Krog, N., Riisom, T.H. & Larsson, K. in *Encyclopedia of Emulsion*88*Technology* (ed. Becker, P.) p. 321 (Marcel Dekker, New York, 1985).
Kubow, S. *Trends Food Sci. Technol.* **1**, 67 (1990).
Kuhn, H. & Möbius, D. *Angew. Chem.* **83**, 672 (1971).
Kyle, D. *Lipid Technol.* **4**, 59 (1992).
Landh, T. *J. Phys. Chem.* in the press (1993).
Landh, T. (Thesis, University of Lund, Sweden), to be published.
Lane, N.J. & Harrison, J.B. *J. Cell. Biol.* **39**, 353 (1979).
Langmuir, I. *J. Amer. Chem. Soc.* **39**, 1848 (1917).
Larsson, K. *Acta Cryst.* **16**, 741 (1963a).
Larsson, K. *Proc. Chem. Soc.* 87 (1963b).
Larsson, K. *Arkiv Kemi* **23**, 35 (1964a).
Larsson, K. *Arkiv Kemi* **23**, 1 (1964b).
Larsson, K. *Acta Cryst.* **21**, 267 (1966a).
Larsson, K. *J. Am. Oil Chem. Soc.* **43**, 559 (1966b).
Larsson, K. *Acta Chem. Scand.* **20**, 2255 (1966c).
Larsson, K. *Nature* **213**, 383 (1967a).
Larsson, K. *Z. Phys. Chem.* **56**, 173 (1967b).
Larsson, K. in *Surface Active Lipids in Foods* p. 8 (Society of Chemical Industry, London, 1968).
Larsson, K. *Fette Seifen Anstrichm.* **74**, 136 (1972).
Larsson, K. *Biochim. Biophys. Acta* **318**, 1 (1973a).
Larsson, K. in *Surface and Colloid Science* (ed. Matijevic, E.) p. 261 (J. Wiley and Sons, New York, 1973b).
Larsson, K. *Chem. Phys. Lipids* **20**, 225 (1977).
Larsson, K. *Prog. Chem. Fats other Lipids* **16**, 163 (1978).
Larsson, K. *J. Colloid Interface Sci.* **72**, 152 (1979).
Larsson, K. *Nature* **304**, 664 (1983).
Larsson, K. *J. Colloid Interface Sci.* **113**, 299 (1986).
Larsson, K. in *The Lipid Handbook* (eds. Gunstone, F.D., Harwood, J.L. & Padley, F.B.) p. 321 (Chapman And Hall, London, 1986).
Larsson, K. *Chem. Phys. Lipids* **49**, 65 (1988).
Larsson, K. *J. Phys. Chem.* **93**, 7304 (1989).
Larsson, K. *J. Am. Oil Chem. Soc.* **69**, 835 (1992a).
Larsson, K. in *Emulsions - A Fundamental and Practical Approach* (ed. Sjöblom, J.) p. 41 (Kluwer Academic Publ., London, 1992b).
Larsson, K. in *Organized Solutions* (eds. Friberg, S. & Lindman, B.) p. 249 (Marcel Dekker, New York, 1992c).

Larsson, K. *J. Dispersion Sci. Technol.* in the press (1994).
Larsson, K. & Al-Mamun, M.A. *Chem. Phys. Lipids* **12**, 176 (1974).
Larsson, K. & Andersson, S. *Acta Chem. Scand.* **A60**, 313 (1986).
Larsson, K., Björnberg, A. & Hellgren, L. *Opscula Medica* **4**, 162 (1975).
Larsson, K., Fontell, K. & Krog, N. *Chem. Phys. Lipids* **27**, 321 (1980).
Larsson, K. & Johansson, L.-Å. *Lebensm.-wiss. Technol.* **11**, 206 (1978).
Larsson, K. & Krog, N. *Chem. Phys. Lipids* **10**, 177 (1973).
Larsson, K. & Lindblom, G. *J. Dispersion Sci. Technol.* **3**, 61 (1982).
Larsson, K. & Lundström, I. in *Lytropic Liquid Crystals* (Adv. in Chem. Series 152) (ed. Friberg, S.) p. 43 (ACS, Washington, 1976).
Larsson, K. & Miezis, Y. To be published (1994).
Larsson, K. & Odham, G. *Marine Pollution Bulletin* **1**, 122 (1970).
Larsson, K. & Puang-Ngern, S. in *Advances in the Biochemistry and Physiology of Plant Lipids* (eds. Appelqvist, L.-Å. & Liljenberg, C.) p. 27 (Elsevier, Amsterdam, 1979).
Larsson, K. & Rand, R.P. *Biochim. Biophys. Acta* **326**, 245 (1973).
Laughlin, R.G., Munyon, R.L., Fu, Y.-C. & Fehl, A.J. *J. Phys. Chem.* **94**, 2546 (1990).
Lewis, R.N.A.H., Mak, N. & Edhaney, R.N.M. *Biochemistry* **26**, 6118 (1987).
Lindblom, G., Larsson, K., Johansson, L., Fontell, K. & Forsén, S. *J. Am. Chem. Soc.* **101**, 5465 (1979).
Lindblom, G. & Rilfors, L. *Biochim. Biophys. Acta* **988**, 221 (1989).
Lindstedt, M., Thompson, R.A. & Allenmark, S. *J. Colloid Interface Sci.* in the press (1993).
Lindström, M., Ljusberg-Wahren, H., Larsson, K. & Borgström, B. *Lipids* **16**, 749 (1981).
Lingwood, C.A., Law, H., Pelizzari, A., Shermans, P. & Drumm, B. *Lancet* **2**, 38 (1989).
Litchfield, C. & Greenberg, A.J. *Comp. Biochem. Physiol.* **47B**, 401 (1974).
Loomis, C.R., Shipley, G.G. & Small, D.M. *J. Lipid Res.* **20**, 525 (1979).
Lösche, M., Duwe, H. & Möhwald, H. *Colloid Interface Sci.* **126**, 432 (1988).
Lundqvist, M. *Prog. Chem. Fats other Lipids* **16**, 101 (1978).
Luzzati, V., Mustacchi, H., Skoulios, A. & Husson, F. *Acta Cryst.* **13**, 660 (1960).
Luzzati, V., Tardieu, A., Gulik-Krzywicki, T. & Rivas, E. *Nature* **220**, 485 (1968).
Mahon, A.J.M. in *Emulsions - A Fundamental and Practical Approach* (ed. Sjöblom, J.) p. 135 (Kluwer Academic Publ., Amsterdam, 1992).
Maltice, W.L., Riiser, J.M. & Clark, D.S. *Biochemistry* **15**, 4272 (1976).
Marankao, R.C., Tercyakz, A.M. & Redgrave, T.G. *Biochim. Biophys. Acta* **875**, 247 (1986).
Mariani, P., Luzzati, V. & Delacroix, H. *J. Mol. Biol.* **204**, 165 (1988).
Matsuno, R. & Adachi, S. *Trends Food Sci. Technol.* **4**, 256 (1993).
Menou, V.B. & Wasan, D.T. in *Encyclopedia of Emulsion Technology* (eds. Becker, P.) p. 1 (Marcel Dekker, New York, 1985).
Menzel, A. & Perco, H. *Wien Med. Wschr.* **19**, 517 (1869).
Miller, K.J., Rao, Y.K., Goodwin, S.R. & Shah, D.O. in *Prediction of Percutaneous Penetration Proceedings* (ed. Hadgraft, J.) (IBC Technical Services, London, 1992).
Minifie, B.W. *Chocolate, Cocoa and Confectionery* (J. & A. Churchill, London, 1970).
Mizukoshi, M. in *Food Emulsions* (eds. Larsson, K. & Friberg, S.) p. 479 (Marcel Dekker, New York, 1989).
Moberger, L. & Larsson, K. *J. Dispersion Sci. Technol.* **6**, 383 (1985).
Moberger, L., Larsson, K., Buchheim, W. & Timmen, H. *J. Dispersion Sci. Technol.* **8**, 207 (1987).
Mohamed, H.M.A. & Larsson, K. *Fat Sci. Technol.* **94**, 338 (1992).
Mohamed, H.M.A. & Larsson, K. *Fat Sci. Technol.* **96**, 57 (1994).
Moore, B.G., Knobler, C.M., Broseta, D. & Rondelez, *J. Chem. Soc. Faraday Trans. II* **82**, 1753 (1987).
Muller, A. *Proc. Roy. Soc.* **A127**, 417 (1930).
Murphy, D.J. *INFORM* **4**, 992 (1993).
Muuse, B.G., Cuperus, F.P. & Derksen, J.T.P. *Ind. Crops and Products* **1**, 57 (1992).
Neveu, D.M. de, Rand, R.P., Ginger, D. & Parsegian, A. *Science* **191**, 399 (1976).
Nichols, B. *Lipid Technol.* **5**, 57 (1993).
Nilsson, P.-G., Lindman, B. & Laughlin, R.G. *J. Phys. Chem.* **88**, 6357 (1984).
Nolte, R.T. & Atkinson, D. *Biophys. J.* **63**, 1221 (1992).
Norling, T., Lading, P., Engström, S., Larsson, K., Krog, N. & SQe Nissen, S. *J. Clin. Periodontol.* **19**, 687 (1992).
Oh, S.G., Klein, S.P. & Shah, D.O. *AICAE Journal* **38**, 149 (1992).

Oh, S.G. & Shah, D.O. *Langmuir* **7**, 1316 (1991).
Oh, S.G., Singh, P. & Shah, D.O. *Langmuir* **8**, 2846 (1992).
Olson, N.F., Huut, C.A., Szoka, F., Vail, W.I. & Papahadjopoulis, D. *Biochim. Biophys. Acta* **557**, 9 (1978).
Öhman, P. *Acta Zool.* **55**, 245 (1974).
Paolo, N.D., Bouncristiani, U., Capotondo, L., Gaggioti, E., DeMia, M., Rossi, P., Sansoni, E. & Bernini, M. *Nephron* **44**, 365 (1986).
Parsegian, V.A. & Rand, R.P. *Langmuir* **7**, 1299 (1991).
Pascher, I. & Sundell, S. *Chem. Phys. Lipids* **20**, 175 (1977).
Pascher, I. & Sundell, S. *Biochim. Biophys. Acta* **855**, 68 (1985).
Pascher, I., Sundell, S. & Hauser, H. *J. Mol. Biol.* **153**, 791 (1981).
Patton, J.S. & Carey, M.C. *Science* **204**, 145 (1979).
Pearson, R. & Pascher, I. *Nature* **281**, 499 (1979).
Pilman, E., Tornberg, E. & Larsson, K. *J. Dispersion Sci. Technol.* **3**, 335 (1982).
Potts, R.O. & Francoeur, M.L. in *Pharmaceutical Skin Penetration88Enhancement* (eds. Walters, K.A. & Hadgraft, J.) p. 269 (Marcel Dekker, New York, 1993).
Prost, J. & Roudelez, F. *Nature* **350**, 11 (1991).
Rajah, K. *Lipid Technol.* **4**, 129 (1992).
Rand, R.P. *Biochim. Biophys. Acta* 241, 823 (1971).
Rand, R.P. in *Food Emulsions* (ed. Friberg, S.) p. 277 (Marcel Dekker, New York, 1976).
Redgrave, T.G., Vassilio, G.G. & Callom, M.J. *Biochim. Biophys. Acta* **1167**, 67 (1992).
Riiner, U. *Lebensm.-W. Technol.* **4**, 81 (1971).
Riisom, T., Krog, N. & Eriksen, J. *J. Cereal Sci.* **2**, 105 (1984).
Ritchie, F.M. *J. Coll. Interface Sci.* **105**, 119 (1985).
Roediger, W.E.W. *Lancet* **2**, 712 (1980).
Roediger, W.E.W. *Gastroenterology* **83**, 424 (1981).
Ross, R. *Nature* **362**, 801 (1993).
Rossell, B.J. *Lipid Technol.* **4**, 106 (1992).
Sada, E., Katoh, S., Tereshima, M., Kheirolomoom, A. & Sawai, H. *Eng. Japan* **20**, 531 (1987).
Sadoc, J.F. & Charvolin, J. *J. Phys. (France)* **47**, 683 (1986).
Sato, K., Arishima, T., Wang, Z.H., Ojima, K., Sagi, N. & Mori, H. *J. Am. Oil Chem. Soc.* **66**, 664 (1989).
Seddon, J.M. *Biochim. Biophys. Acta* **1031**, 1 (1990).
Seddon, J.M., Cevc, G., Kayl, R.D. & Marsh, D. *Biochemistry* **23**, 2634 (1984).
Sefton, J., Arnebrant, T. & Glantz, P.-O. *Acta Odontol. Scand.* **50**, 221 (1992).
Shepherd, I.S. & Yoell, R.W. in *Food Emulsions* (ed. Friberg, S.) p. 215 (Marcel Dekker, New York, 1976).
Shipley, G.G., Green, J.P. & Nichols, B.W. *Biochim. Biophys. Acta* **311**, 531 (1973).
Shipley, G.G., Avecilla, L.S. & Small, D.M. *Biochemistry* **24**, 2902 (1985).
Simpson, T.D. & Hagemann, J.W. *J. Am. Oil Chem. Soc.* **59**, 169 (1982).
Singer, S.J. & Nicholson, G.L. *Science* **175**, 720 (1972).
Slade, L. & Levine, H. *Carbohydrate Polymers* **8**, 183 (1988).
Slomiany, B.L., Murty, V.L.M. & Slomiany, A. *Prog. Lipid. Res.* **24**, 311 (1985).
Small, D.M. *Handbook of Lipid Research* 4. *The Physical Chemistry of Lipids. From Alkanes to Phospholipids.* p. 511 (Plenum Press, New York, 1986).
Smith, L.L. *Chem. Phys. Lipids* **44**, 87 (1987).
Söderberg, I. *Structural Properties of Monoglycerides, Phospholipids and Fats in Aqueous systems* (Thesis, University of Lund, 1990).
Söderberg, I., Hernqvist, L. & Buchheim, W. *Milchwissenschaft* **44**, 403 (1989).
Sripada, P.K., Malik, P.R., Hamilton, J.A. & Shipley, G.G. *J. Lipid Res.* **28**, 710 (1987).
Stavans, J. & Glazier, J.A. *Phys. Rev. Lett.* **A37**, 1318 (1989).
Sternberg, B., Gumpert, J., Reinhart, G. & Gawrisch, K. *Biochim. Biophys. Acta* **898**, 223 (1987).
Stremmel, W. *J. Clin. Invest.* **82**, 2001 (1988).
Subramaniam, S. & Connel, H.M.M. *J. Phys. Chem.* **91**, 1715 (1987).
Suzuki, M., Ogaki, T. & Sato, K. *J. Am. Oil. Chem. Soc.* **62**, 1600 (1985).
Svensson, I., Adlercreutz, P., Mattiasson, B., Miezis, Y. & Larsson, K. *Chem. Phys. Lipids* in the press (1993).
Swarm, R.A., Ashle, S.W., Soybel, D.I., Ordway, F.S. & Cheung, L.Y. *Am. J. Surgery* **153**, 48

(1987).
Szoka, F. & Papahadjopoulis, D. *Proc. Nat. Acad. Sci. (USA)* **75**, 4194 (1978).
Szoka, F. & Papahadjopoulis, D. *Amer. Rev. Biophys. Bioeng.* **9**, 467 (1980).
Tanford, C. *The Hydrophobic Effect: Formation of Micelles and Biological Membranes* (Wiley, New York, 1980).
Taylor, P. *Lipid Technol.* **3**, 9 (1991).
Tempel, M.v.D. *J. Colloid Sci.* **16**, 284 (1961).
Tempel, M.v.D. *Surface-Active Lipids in Foods* pp. 1-22 (S.C.I. Monograph, London, 1968).
Terech, P., Volino, F. & Ramasseul, R.J. *J. Physique* **46**, 895 (1985).
Timms, R.E. *Prog. Lipid Res.* **23**, 1 (1984).
Titball, R.W. *Microbiol. Rev.* **57**, 347 (1993).
Tornberg, E., Olsson, A. & Persson, K. in *Food Emulsions* (eds. Larsson, K. & Friberg, S.) p. 247 (Marcel Dekker, New York, 1989).
Trommelen, A.M. in *Proc. 17th Nordic Lipid Symposium* (eds. Lambertsen, G. & Mlkki, Y.) p. 18 (Lipidforum, N-5035 Bergen, Norway, 1993).
Tscharner, V.V. & Connel, H.M. *Biophys. J.* **36**, 409 (1981).
Untersee, P., Gil, J. & Weibel, E.R. *Respirat. Physiol.*, **13**, 171 (1971).
Vand, V., Lomer, T.R. & Lang, A. *Acta Cryst.* **2**, 214 (1949).
Vargas, R., Mariani, P., Gulik, A. & Luzzati, V. *J. Mol. Biol.* **225**, 137 (1992).
von Schnering, H.G. & Nesper, R. *Z. Phys.* **B83**, 407 (1991).
Vulfsen, E.N. *Trends Food Sci. & Technol.* **4**, 209 (1993).
Wahlgren, M. & Arnebrant, T., *Trends Biochem.* **9**, 209 (1991).
Walstra, P. in *Food Structure and Behaviour* (ed. Blanshard, J.M.V.) p. 1 (Academic Press, London, 1987).
Walstra, P. *Chem. Engin. Sci.* **48**, 333 (1993).
Walstra, P. & Jenners, R. *Dairy Chemistry and Physics* (J. Wiley & Sons, New York, 1984).
Walstra, P. & Lillford, P. in *Food Structure and Behaviour* (ed. Blanshard, J.M.V.) p. 67 (Academic Press, London, 1987).
Watanabe, A., Tashima, I., Matsuzaki, M., Kurashige, J. & Sato, K. *J. Am. Oil Chem. Soc.* **69**, 1077 (1982).
Weis, R.M. *Chem. Phys. Lipids* **57**, 227 (1991).
Wertz, P.W. & Downing, D.T. *J. Lipid Res.* **24**, 759 (1983).
Whitburn, K.D. & Dunne, C.P. in *Microemulsions and Emulsions in Foods* (eds. El-Nokaly, M. & Cornell, D.) p. 161 (ACS, Washington D. C., 1991).
Wieslander, Å., Ulmius, J., Lindblom, G. & Fontell, K. *Biochim. Biophys. Acta* **512**, 241 (1978).
William, R.E. *Science* **161**, 276 (1968).
Willie, R.L. & Lutton, E.S. *J. Am. Chem. Soc.* **43**, 491 (1966).
Winkler, F., D'Arcy, A. & Hunziker, W. *Nature* **343**, 771 (1990).
Yasukawa, T. & Aigami, K. *Jap. Soc. Food Eng.* **11**, 3 (1989).
Yoshimoto, N., Nakamura, T., Suzuki, M. & Sato, K. *J. Phys. Chem.* **95**, 3384 (1991).
Zilversmit, D.B. *J. Biol. Chem.* **249**, 2645 (1971).

List of Abbreviations

BP I,II & III	blue phases
CBA	cocoa butter alternatives
CBE	cocoa butter equivalents
CBS	cocoa butter substitutes
C_D-phase	cubic diamond phase
C_G-phase	cubic gyroid phase
CMC	critical micellar concentration
C_P-phase	cubic primitive phase
CSL	calcium salts of stearoyl lactylates
DATEM	diacetyl-tartaric esters of monoglycerides
DGDG	digalactosyldiglyceride or digalactosyldiacylglycerol
DLVO	Deryaguin-Landau-Verwey-Overbeck
DSC	differential scanning calorimetry
DPPC	dipalmitoylphosphatidylcholine
EHEC	ethyl-hydroxyethyl-cellulose
ESP	equilibrium spreading pressure
G	gaseous monolayer
HDL	high-density lipoproteins
IPMS	infinite periodic minimal surface
LB	Langmuir-Blodgett (film)
LC	liquid-condensed (monolayer)
LDL	low-density lipoprotein
LE	liquid-expanded monolayer
M	myristic acid
MGDG	monogalactosyldiglyceride or monogalactosyldiacylglycerol
O	oleic acid
O/W	oil-in-water
O/W/O	oil-in-water-in-oil
P	palmitic acid
PA	phosphatidic acid
PC	phosphatidylcholine
PE	phosphatidylethanolamine
PG	phosphatidylglycerol
PI	phosphatidylinositol
PS	phosphatidylserine
RES	reticulo-endothelial system
S	solid
S or St	stearic acid
SAX	small-angle X-ray diffraction
SCFA	short-chain fatty acid
s-CO_2	supercritical carbon dioxide
SDS	sodium dodecylsulphate
SM	sphingomyelin
SSL	sodium salts of stearoyl lactylates
VLDL	very-low density lipoprotein
W/O	water-in-oil
W/O/W	water-in-oil-in-water
WSB	water-saturated butanol

INDEX